Table of Contents

1 Chapter 1: Introduction to Membrane Biotics

Vision is a complex process, more than just light entering the eye. It's a cascade of events, starting with photons hitting the retina and culminating in our brain interpreting a visual scene. When things go wrong in this intricate system, vision impairments arise. From the common refractive errors like nearsightedness and farsightedness to more severe conditions like macular degeneration or retinitis pigmentosa, these impairments significantly impact quality of life. Understanding the root causes of these problems is the first step towards developing effective solutions, which is where the exciting field of membrane biotics comes in.

At its core, membrane biotics focuses on manipulating cellular membranes, especially those in the eye, to improve visual function. Cellular membranes are not static barriers; they are dynamic structures crucial for cell signaling, nutrient transport, and overall cellular health. In the context of vision, the membranes of photoreceptor cells – the rods and cones in the retina – are of paramount importance. These membranes contain the light-sensitive pigments that initiate the visual cascade. Dysfunction in these membranes, whether due to genetic mutations, aging, or environmental factors, can lead to impaired vision.

Therefore, by understanding the composition and function of these membranes, and by developing methods to modify them, we can potentially restore or enhance visual capabilities. This approach moves beyond traditional treatments that simply correct optical errors and ventures into directly influencing the biological processes underlying vision.

Now, to truly grasp the potential of membrane biotics, we need a solid foundation in optical principles. Light, as both a wave and a particle, behaves in predictable ways when it interacts with matter. The eye itself is an optical instrument, designed to focus light onto the retina, creating a clear image. This focusing process relies on refraction, the bending of light as it passes from one medium to another. The cornea and the lens, two key structures in the eye, are responsible for most of this refraction. The shape of these structures determines how effectively light is focused, and imperfections in their shape can lead to refractive errors. Myopia (nearsightedness) occurs when the eye is too long or the cornea is too curved, causing light to focus in front of the retina. Hyperopia (farsightedness) is the opposite, where the eye is too short or the cornea is too flat, causing light to focus behind the retina. Astigmatism arises from an irregularly shaped cornea, leading to distorted vision at all distances. Corrective lenses, whether glasses or contact lenses, compensate for these refractive errors by providing the additional refraction needed to focus light properly on the retina. Furthermore, it is also important to consider other optical phenomena. Diffraction, the bending of light around obstacles, and interference, the combination of light waves to create constructive or destructive patterns, also play a role in image formation in the eye. Aberrations, deviations from perfect image formation, can arise from imperfections in the optical system of the eye. These aberrations can blur vision, reduce contrast sensitivity, and cause halos around lights. While corrective lenses can address some aberrations, more advanced techniques, such as wavefront-guided laser vision correction, can correct even more complex distortions. Understanding these optical principles is essential for understanding how the eye works, and for developing strategies to correct vision impairments.

Membrane biotics offers a new perspective. Imagine that, instead of solely relying on external lenses to correct refractive errors, we could directly modify the membranes of the cells responsible for vision. What if we could strengthen the membranes of photoreceptor cells, making them more resistant to age-related damage? What if we could enhance the light-sensitive pigments in these membranes, improving visual acuity? These are the kinds of possibilities that membrane biotics explores.

The field requires a multidisciplinary approach, bringing together expertise in ophthalmology, optical physics, molecular biology, and materials science. By combining our knowledge of these fields, we can develop novel strategies to fine-tune the cellular mechanisms underlying vision. The journey begins with a deep understanding of the structure and function of cellular membranes and the optical properties of the eye. It then progresses to exploring ways to manipulate these membranes, ultimately leading to new and improved treatments for a wide range of vision impairments. It is an ambitious undertaking, but one with the potential to transform the lives of millions of people.

1.1 Vision Impairments Overview

Vision impairments encompass a wide spectrum of conditions, each affecting visual acuity, field of vision, or color perception in distinct ways. Refractive errors, such as myopia (nearsightedness), hyperopia (farsightedness), and astigmatism, represent the most common category. Myopia, often developing in childhood, arises when the eye's shape causes light to focus in front of the retina instead of directly on it, resulting in blurred distance vision. Conversely, hyperopia occurs when light focuses behind the retina, leading to blurry near vision, although distant objects may also appear unclear. Astigmatism results from an irregularly shaped cornea or lens, causing light to focus unevenly on the retina, producing distorted or blurred vision at all distances. These errors are generally correctable with eyeglasses, contact lenses, or refractive surgery, yet understanding the underlying anatomical variations is crucial for comprehending membrane biotics applications.

Moving beyond refractive errors, we encounter a diverse group of conditions affecting various parts of the eye. Cataracts, characterized by clouding of the eye's natural lens, are a leading cause of blindness worldwide. The gradual opacification of the lens obstructs light passage, resulting in blurred vision, glare, and difficulty seeing in low light conditions. Age-related macular degeneration (AMD) affects the macula, the central part of the retina responsible for sharp, detailed vision. In its dry form, AMD involves the gradual thinning of the macula, leading to a slow decline in central vision. The wet form, more aggressive, involves the growth of abnormal blood vessels under the retina, which leak fluid and blood, causing rapid vision loss. Glaucoma, often associated with elevated intraocular pressure, damages the optic nerve, which transmits visual information from the eye to the brain. This damage typically starts in the peripheral vision, gradually progressing towards central vision, often without noticeable symptoms in the early stages.

Retinal disorders represent another significant category of vision impairments. Diabetic retinopathy, a complication of diabetes, damages the blood vessels in the retina, leading to leakage, swelling, and the formation of abnormal new vessels. This can cause blurred vision, floaters, and even blindness. Retinitis pigmentosa (RP) is a group of inherited disorders that cause progressive degeneration of the photoreceptor cells in the retina, leading to night blindness and gradual loss of peripheral vision, eventually resulting in tunnel vision. Finally, color vision deficiencies, often inherited, result from the absence or dysfunction of one or more types of cone cells in the retina, which are responsible for color perception. The most common form is red-green color blindness, where individuals have difficulty distinguishing between red and green hues.

Furthermore, the intricacies of the cellular membranes within the eye, particularly those of photoreceptor cells, play a critical role in the pathophysiology of these conditions. In AMD, for example, the dysfunction of the retinal pigment epithelium (RPE), a layer of cells supporting the photoreceptors, leads to the accumulation of cellular debris and oxidative stress, ultimately damaging the photoreceptor membranes. Similarly, in glaucoma, the structural integrity of the optic nerve fibers, including their cellular membranes, is compromised by elevated pressure and other factors, leading to axonal damage and cell death. In diabetic retinopathy, the damage to the blood vessels in the retina disrupts the delicate balance of oxygen and nutrients, leading to oxidative stress and damage to the photoreceptor membranes. Understanding the specific mechanisms by which these conditions affect cellular membranes is crucial for developing targeted therapeutic interventions. For instance, in RP, gene therapies aimed at restoring the function of mutated genes involved in photoreceptor membrane protein synthesis and trafficking hold promise. Similarly, in AMD, therapies aimed at reducing oxidative stress and inflammation in the RPE and photoreceptor membranes may help slow down the progression of the disease. By targeting the underlying cellular and molecular mechanisms that contribute to membrane dysfunction in these conditions, we can potentially develop more effective and personalized treatments for vision impairments. The subsequent chapters will delve into how manipulating membrane composition and function can offer novel strategies for visual restoration and enhancement, particularly in cases where conventional treatments prove insufficient.

It's important to note that, beyond these common conditions, various other factors can contribute to vision impairment, including infections, injuries, and neurological disorders. Infections such as conjunctivitis and keratitis can cause inflammation and damage to the cornea and conjunctiva, leading to temporary or permanent vision loss. Traumatic injuries to the eye, such as corneal abrasions, penetrating injuries, and orbital fractures, can also result in significant vision impairment. Neurological disorders, such as stroke, multiple sclerosis, and brain tumors, can affect the visual pathways in the brain, leading to a variety of visual disturbances, including double vision, visual field loss, and cortical blindness. Therefore, a comprehensive understanding of vision impairments requires considering a wide range of potential causes and mechanisms, with a focus on the underlying cellular and molecular processes that contribute to visual dysfunction.

1.2 Optical Mechanisms Primer

Light, the very essence of sight, interacts with our eyes through a series of intricate optical mechanisms. Understanding these mechanisms is paramount to grasping how Membrane Biotics aims to fine-tune vision impairments. These processes involve the bending, focusing, and detection of light, all orchestrated by the eye's sophisticated anatomy. This is the first step in understanding where things can go wrong, and what the path could be to repairing them.

The journey of light begins as it passes through the cornea, the transparent outer layer of the eye. Its curved shape is the first and most powerful refractive surface in the visual system. This curvature bends light rays, initiating the focusing process. Think of it like a lens on a camera, starting to converge the incoming light towards a single point. Any irregularities in the corneal surface, such as those seen in astigmatism, will distort the light and blur vision, it is vital to understand these principles, so that the causes and potential corrections can be well thought out.

After the cornea, light encounters the aqueous humor, a clear, watery fluid filling the space between the cornea and the lens. This fluid maintains intraocular pressure, keeping the eyeball's shape and providing nutrients. While the aqueous humor contributes minimally to overall refraction, its clarity is crucial for unobstructed light passage. From there, the light enters the pupil, the adjustable aperture controlled by the iris, which modulates the amount of light entering the eye, similar to the diaphragm of a camera, adjusting to varying light conditions. In bright light, the iris constricts the pupil, reducing light intake; in dim light, it dilates the pupil, allowing more light in. This adaptation helps to optimize image quality and protect the retina from excessive light exposure.

The lens, located behind the iris, provides further refinement in focusing. Unlike the cornea, the lens is dynamic, capable of changing its shape to focus on objects at varying distances. This process, known as accommodation, is achieved through the action of the ciliary muscles, which adjust the tension on the zonular fibers that suspend the lens. When focusing on distant objects, the ciliary muscles relax, flattening the lens; when focusing on near objects, the ciliary muscles contract, thickening the lens. This precise adjustment ensures that the image is sharply focused on the retina. However, as we age, the lens loses its elasticity, making accommodation more difficult, leading to presbyopia, the age-related difficulty in focusing on near objects.

Following the lens, light travels through the vitreous humor, a clear, gel-like substance that fills the majority of the eye's volume, maintaining its spherical shape and supporting the retina. The vitreous humor must be transparent to allow light to reach the retina without distortion. Opacities or debris in the vitreous humor, known as floaters, can cast shadows on the retina, appearing as small specks or strands in the field of vision, thus any irregularities or changes in transparency can affect vision clarity.

Finally, light reaches the retina, a light-sensitive layer lining the back of the eye. The retina contains photoreceptor cells, rods and cones, which convert light into electrical signals that are transmitted to the brain via the optic nerve. Rods are responsible for vision in low light conditions and are highly sensitive to motion and peripheral vision, whilst Cones are responsible for color vision and visual acuity in brighter conditions. They are concentrated in the macula, the central part of the retina, and are essential for detailed vision, like reading and driving. The distribution and function of these photoreceptor cells are vital for proper visual perception, and any damage or dysfunction can lead to significant vision impairments.

Beyond the basic refraction and focusing, the eye also deals with imperfections known as aberrations. These aberrations distort the image formed on the retina, degrading visual acuity. Aberrations can be classified as lower-order aberrations, such as myopia (nearsightedness), hyperopia (farsightedness), and astigmatism, or higher-order aberrations, which are more complex distortions. While lower-order aberrations can be corrected with glasses or contact lenses, higher-order aberrations are more challenging to address and often require advanced corrective procedures.

The eye's optical system is also subject to scattering, where light is deflected in various directions due to particles or irregularities in the ocular media. Scattering reduces image contrast and sharpness, particularly in conditions like cataracts, where the lens becomes cloudy. Minimizing scattering is crucial for maintaining clear and high-quality vision. Therefore, understanding how light interacts with these components is imperative to find ways to restore clarity.

The eye, although a marvel of biological engineering, is not without its limitations. Imperfections in the cornea, lens, or other ocular structures can lead to various vision impairments. The goal of Membrane Biotics is to address these imperfections by manipulating cellular membranes to enhance the optical properties of the eye and improve visual function. Understanding the intricate interplay of these optical mechanisms and their potential for disruption is essential for developing effective strategies to fine-tune human vision impairments with Membrane Biotics. It's not just about seeing; it's about understanding how we see, and how we can see better.

2 Chapter 2: Cellular Membranes and Vision

Cellular membranes aren't just static barriers; they are dynamic interfaces, crucial for every aspect of cell function, and particularly relevant to vision. The lipid bilayer, primarily composed of phospholipids, forms the basic structure, but the diversity of lipids and the presence of various proteins embedded within it are what dictate its functionality. Think of the membrane as a bustling city street, where lipids are the pavement and proteins are the buildings, each playing a specific role in the city's operations. Cholesterol, another important lipid component, modulates membrane fluidity. At high temperatures, it prevents the membrane from becoming too fluid, while at low temperatures, it prevents it from solidifying. This delicate balance is essential for proper membrane function.

In the context of vision, the photoreceptor cells in the retina, specifically rods and cones, heavily rely on the unique properties of their cellular membranes. The outer segment of these cells, where phototransduction occurs, is characterized by a stack of flattened, disc-like structures formed by specialized membranes. These discs house rhodopsin (in rods) and cone opsins, the light-sensitive proteins responsible for capturing photons and initiating the visual cascade. The formation, maintenance, and turnover of these discs are critical processes that directly impact visual sensitivity and overall retinal health. Furthermore, the phospholipid composition of the photoreceptor membrane is highly enriched in polyunsaturated fatty acids (PUFAs), particularly docosahexaenoic acid (DHA). DHA contributes to membrane fluidity, which is essential for the rapid diffusion of proteins involved in phototransduction.

Photoreceptor membrane dynamics are a complex interplay of synthesis, transport, and degradation. New disc membranes are continuously synthesized at the base of the outer segment and gradually move towards the tip, where they are eventually shed and phagocytosed by the retinal pigment epithelium (RPE). This continuous turnover ensures that damaged or dysfunctional proteins are removed, maintaining the integrity of the visual system. The rate of disc shedding and phagocytosis is tightly regulated and can be affected by factors such as light exposure, age, and disease. Disruptions in this process can lead to the accumulation of debris and the formation of lipofuscin, a hallmark of age-related macular degeneration (AMD). The proper trafficking of proteins within the photoreceptor membrane is also crucial for their function. Specialized transport mechanisms ensure that rhodopsin and other phototransduction proteins are delivered to the correct location within the disc membrane. This process involves a complex network of chaperones and motor proteins that guide the proteins along microtubule tracks.

Membrane protein interactions within photoreceptor cells are intricate and highly specific. Rhodopsin, for instance, interacts with several other proteins, including transducin, phosphodiesterase (PDE), and rhodopsin kinase, to initiate and regulate the phototransduction cascade. When light strikes rhodopsin, it undergoes a conformational change that allows it to activate transducin, a G protein. Activated transducin then activates PDE, which hydrolyzes cyclic GMP (cGMP), leading to the closure of cGMP-gated ion channels in the plasma membrane. This closure reduces the influx of sodium ions, hyperpolarizing the cell and reducing neurotransmitter release. The interaction between rhodopsin and rhodopsin kinase is crucial for terminating the light response. Rhodopsin kinase phosphorylates rhodopsin, which allows it to bind to arrestin, preventing further activation of transducin. These protein-protein interactions are finely tuned to ensure a rapid and efficient response to light.

Beyond the core phototransduction machinery, other membrane proteins play important roles in maintaining photoreceptor cell health and function. For example, the ATP-binding cassette transporter ABCA4 is essential for removing toxic retinoids from the photoreceptor membrane. Mutations in ABCA4 are a major cause of Stargardt disease, an inherited form of macular degeneration. The dysfunction of ABCA4 leads to the accumulation of lipofuscin and subsequent retinal damage. Furthermore, membrane receptors for various growth factors and signaling molecules are present in photoreceptor cells, allowing them to respond to external cues and modulate their function. These receptors can influence cell survival, differentiation, and even the expression of phototransduction proteins. Understanding these membrane protein interactions and their roles in retinal health is crucial for developing effective therapies for vision impairments. By targeting specific protein interactions or manipulating membrane composition, it may be possible to restore function to damaged photoreceptor cells and prevent further vision loss. Imagine, for instance, designing small molecules that specifically enhance the interaction between rhodopsin and rhodopsin kinase, leading to a more rapid termination of the light response and improved visual acuity in individuals with certain retinal disorders. The possibilities are vast, and further research in this area is essential for unlocking the full potential of membrane biotics in the treatment of vision impairments.

2.1 Membrane Structure & Function

The plasma membrane, a seemingly simple barrier, is in reality a dynamic and complex structure that governs a cell's interactions with its environment. Understanding its architecture and function is paramount to grasping the fundamental processes underpinning vision, especially when considering the photoreceptor cells that initiate the visual cascade. At its core, the plasma membrane is composed of a lipid bilayer, primarily phospholipids, arranged in a manner that creates a hydrophobic core and hydrophilic surfaces. These phospholipids have a polar head group, containing phosphate, and two non-polar fatty acid tails. The arrangement arises from the amphipathic nature of these molecules, meaning they possess both hydrophilic and hydrophobic regions. The tails cluster together to exclude water while the heads interact with the aqueous environment, forming a stable bilayer.

Cholesterol, another crucial lipid component, nestles within the phospholipid bilayer, modulating membrane fluidity and stability. It acts as a buffer, preventing the membrane from becoming too fluid at high temperatures and too rigid at low temperatures. This is particularly important in photoreceptor cells, which must function under a range of physiological conditions. The amount of cholesterol present will dictate the overall rigidity and flexibility of the membrane, which in turn affects the way that proteins can move, how the membrane bends, and whether it can effectively fuse with other structures. The plasma membrane, however, is not just a bag of lipids. Proteins are embedded within or associated with the lipid bilayer, performing a wide range of functions. These proteins can be integral membrane proteins, which span the entire bilayer, or peripheral membrane proteins, which are associated with the membrane surface.

Integral membrane proteins often act as channels, carriers, or receptors. Channels facilitate the selective passage of ions or small molecules across the membrane, while carriers bind to specific molecules and undergo conformational changes to transport them. Receptors bind to signaling molecules, triggering intracellular signaling cascades. In photoreceptor cells, rhodopsin, a light-sensitive receptor protein, is a prime example of an integral membrane protein essential for vision. Peripheral membrane proteins, on the other hand, can be involved in scaffolding, anchoring, or enzymatic activity. They often interact with integral membrane proteins or the cytoskeleton to provide structural support and regulate membrane function. The cytoskeleton, a network of protein filaments, provides structural support to the cell and helps to maintain its shape. It is also involved in cell movement, cell division, and intracellular transport. The cytoskeleton interacts with the plasma membrane through a variety of linker proteins, connecting the membrane to the underlying scaffolding.

Beyond the basic structure, the plasma membrane exhibits lateral heterogeneity, with lipids and proteins clustering into specialized microdomains known as lipid rafts. Lipid rafts are enriched in cholesterol and sphingolipids, creating regions of increased order and rigidity within the fluid lipid bilayer. These rafts serve as platforms for organizing membrane proteins, facilitating protein-protein interactions, and regulating signaling pathways. In photoreceptor cells, lipid rafts play a role in the localization and function of visual transduction proteins. The membrane is far from static; it is highly dynamic. Lipids and proteins are constantly moving laterally within the bilayer, allowing for rapid reorganization and adaptation to changing conditions. This fluidity is essential for many membrane functions, including cell signaling, membrane fusion, and protein trafficking.

The primary function of the plasma membrane is to provide a selective barrier between the intracellular and extracellular environments. It regulates the passage of molecules into and out of the cell, maintaining a stable internal environment. Small, nonpolar molecules can diffuse across the membrane readily, while larger, polar molecules and ions require the assistance of membrane proteins. Transport across the membrane can be passive, driven by concentration gradients, or active, requiring energy input. Passive transport includes simple diffusion, facilitated diffusion (mediated by channels or carriers), and osmosis. Active transport utilizes energy, typically in the form of ATP hydrolysis, to move molecules against their concentration gradients. In photoreceptor cells, ion channels and pumps are essential for maintaining the electrochemical gradients that drive the visual signal.

Another critical function of the plasma membrane is cell signaling. Receptors on the cell surface bind to signaling molecules, triggering intracellular signaling cascades that regulate a variety of cellular processes. These signaling cascades often involve a series of protein-protein interactions, phosphorylation events, and changes in gene expression. In photoreceptor cells, the binding of light to rhodopsin initiates a signaling cascade that ultimately leads to the closing of ion channels and the generation of an electrical signal. The membrane also plays a crucial role in cell-cell communication. Cells can communicate with each other through direct contact, gap junctions, or the release of signaling molecules. Direct contact involves the interaction of membrane proteins on adjacent cells, while gap junctions allow for the direct passage of small molecules and ions between cells. The release of signaling molecules, such as hormones or neurotransmitters, allows cells to communicate over long distances.

Furthermore, the plasma membrane is involved in endocytosis and exocytosis, processes that allow cells to take up and release large molecules or particles. Endocytosis involves the invagination of the plasma membrane to form a vesicle that engulfs extracellular material. Exocytosis involves the fusion of a vesicle with the plasma membrane to release its contents into the extracellular space. These processes are essential for nutrient uptake, waste removal, and the secretion of signaling molecules. In photoreceptor cells, endocytosis and exocytosis are involved in the turnover of visual pigments and the release of neurotransmitters. Finally, the intricate interplay between the membrane's structural components and functional proteins is critical for the health and proper function of the human visual system, a complex system that relies on the precise orchestration of molecular events.

2.2 Photoreceptor Membrane Dynamics

The photoreceptor membrane, a marvel of biological engineering, is the site of the initial events in vision, where light energy is transduced into a neural signal. Understanding the dynamics of this membrane is crucial to unraveling the complexities of visual function and developing effective strategies for treating vision impairments. In essence, the photoreceptor membrane isn't static; it's a highly dynamic structure constantly undergoing remodeling, protein trafficking, and lipid turnover, all of which are essential for maintaining its light-sensing capability.

The outer segment of photoreceptor cells, where visual transduction occurs, is packed with membranous discs. These discs are not directly connected to the plasma membrane in rods, whereas they are continuous with the plasma membrane in cones. Each disc is a lipid bilayer enriched with rhodopsin (in rods) or cone opsins (in cones), the light-sensitive proteins that initiate the visual cascade. The continuous renewal of these discs is a critical process. New discs are formed at the base of the outer segment, and older discs are shed from the tip, a process known as disc shedding. This constant turnover ensures that damaged proteins and lipids are removed, maintaining the efficiency of light capture. The rate of disc turnover varies depending on the species and the type of photoreceptor, but it's a continuous process vital for photoreceptor health. Disruptions in disc turnover have been implicated in retinal degenerative diseases.

Lipid composition profoundly influences membrane dynamics. The photoreceptor membrane is exceptionally rich in polyunsaturated fatty acids (PUFAs), particularly docosahexaenoic acid (DHA). DHA enhances membrane fluidity, which is believed to be essential for the efficient diffusion of phototransduction proteins within the membrane. The distribution of lipids isn't uniform; there are lipid rafts, specialized microdomains enriched in cholesterol and sphingolipids. These rafts serve as platforms for concentrating specific proteins, thereby facilitating protein-protein interactions critical for phototransduction.

Moreover, the asymmetric distribution of phospholipids between the inner and outer leaflets of the membrane plays a crucial role in maintaining membrane integrity and regulating protein function. The enzyme flippase maintains this asymmetry by actively transporting certain phospholipids from one leaflet to the other.

The trafficking and localization of proteins within the photoreceptor membrane are tightly regulated. Rhodopsin, for example, must be transported from its site of synthesis in the inner segment to the outer segment discs. This transport involves a complex interplay of chaperones, motor proteins, and trafficking vesicles. Once in the outer segment, rhodopsin must be correctly oriented within the disc membrane to function efficiently. Other proteins involved in phototransduction, such as transducin, phosphodiesterase, and arrestin, also undergo specific trafficking patterns to ensure their proper localization and function. Disruptions in protein trafficking can lead to mislocalization of these proteins, impairing phototransduction and contributing to retinal degeneration.

The photoreceptor membrane is not an isolated entity; it interacts extensively with the surrounding cytoplasm and the interphotoreceptor matrix (IPM). The IPM is a complex extracellular space between the photoreceptors and the retinal pigment epithelium (RPE). It contains a variety of proteins and glycosaminoglycans that play a role in photoreceptor function and retinal adhesion. For instance, interphotoreceptor retinoid-binding protein (IRBP) shuttles retinoids between the photoreceptors and the RPE, a process essential for the visual cycle. Furthermore, the IPM facilitates the exchange of nutrients and metabolites between the photoreceptors and the RPE. The RPE is responsible for phagocytosing the shed outer segment discs, a critical step in the disc turnover process.

Regulation of phototransduction involves intricate feedback mechanisms that modulate membrane protein activity. Calcium ions play a central role in these feedback loops. Light-induced activation of rhodopsin leads to a decrease in intracellular calcium concentration, which in turn affects the activity of guanylate cyclase, the enzyme responsible for synthesizing cGMP. Changes in cGMP levels regulate the opening and closing of cyclic nucleotide-gated (CNG) channels, controlling the influx of calcium and sodium ions into the photoreceptor. This feedback loop helps to adapt the photoreceptor to varying light intensities, allowing us to see in both bright and dim conditions. Protein phosphorylation also plays a crucial role in regulating phototransduction. Rhodopsin kinase phosphorylates activated rhodopsin, leading to its binding by arrestin, which quenches rhodopsin's activity and initiates its internalization.

External factors significantly influence the dynamics of the photoreceptor membrane. Light exposure itself can induce changes in membrane lipid composition and protein distribution. Oxidative stress, caused by excessive light exposure or metabolic dysfunction, can damage membrane lipids and proteins, leading to photoreceptor dysfunction. Age-related changes also affect membrane dynamics, with a decline in lipid turnover and an increase in protein aggregation. Genetic mutations in genes encoding membrane proteins or enzymes involved in lipid metabolism can disrupt membrane function and lead to inherited retinal diseases. Furthermore, nutritional deficiencies, particularly a lack of essential fatty acids like DHA, can compromise membrane integrity and impair visual function.

Understanding the intricate dynamics of the photoreceptor membrane opens avenues for therapeutic interventions aimed at treating vision impairments. Strategies include modulating lipid composition through dietary supplementation or targeted drug delivery, enhancing protein trafficking by optimizing chaperone function, and protecting the membrane from oxidative damage with antioxidant therapies. Gene therapy holds promise for correcting genetic defects that disrupt membrane protein function. By targeting the specific molecular mechanisms that govern membrane dynamics, we can develop more effective and personalized treatments for a wide range of retinal diseases.

2.3 Membrane Protein Interactions

Membrane protein interactions are the unsung heroes of visual transduction. These intricate partnerships govern every stage of light detection, from the initial capture of photons to the generation of electrical signals that travel to the brain. Understanding these interactions is paramount if we hope to leverage membrane biotics for vision correction. Their correct assembly, stability, and regulated activity are absolutely vital.

The visual cascade, the series of events that transforms light into neuronal signals, relies heavily on specific protein-protein interactions within the photoreceptor cell membrane. Rhodopsin, the light-sensitive protein, sits embedded in the disc membranes of rod cells. It's not a solitary actor. Upon absorbing a photon, rhodopsin undergoes a conformational change, activating transducin, a heterotrimeric G protein. This activation is mediated by direct physical contact between rhodopsin and transducin, a dance orchestrated by specific amino acid sequences on both proteins. Disrupting this interaction, even subtly, can cripple the entire visual process, which needs to be understood better. We are looking to create tools to fine-tune the interactions for therapeutic purposes, allowing for the development of personalized medicine.

Think of it like a complex switchboard: rhodopsin flips the switch, activating transducin, which in turn activates phosphodiesterase (PDE). PDE hydrolyzes cyclic GMP (cGMP), a nucleotide that keeps ion channels open in the dark. As cGMP levels plummet, these channels close, leading to hyperpolarization of the photoreceptor cell, ultimately sending a signal to the brain. Each of these steps, from transducin activating PDE to PDE's interaction with cGMP-gated channels, is governed by specific protein-protein interactions, each one influencing the other. The binding affinities, the rates of association and dissociation, are all carefully controlled to ensure rapid and efficient signal transduction. Any aberrant interaction can cause severe visual impairments.

Beyond the core visual cascade, numerous other proteins interact with those mentioned above, modulating their activity and ensuring the system doesn't run amok. Arrestin, for example, binds to phosphorylated rhodopsin, effectively turning off the light-activated rhodopsin molecule. This prevents overstimulation and allows the photoreceptor cell to reset for the next photon. Recoverin, a calcium-binding protein, regulates rhodopsin kinase, the enzyme responsible for phosphorylating rhodopsin. These regulatory proteins are critical for maintaining the sensitivity and dynamic range of vision. Without them, we would either be blinded by bright light or unable to see in dim conditions, which clearly shows how we need to be careful when handling our system.

Lipid rafts also play a crucial role. These are specialized microdomains within the membrane enriched in cholesterol and sphingolipids. They serve as platforms for concentrating specific proteins and facilitating their interactions. For instance, rhodopsin and some of its interacting partners, like transducin, tend to cluster within lipid rafts, enhancing the efficiency of signal transduction. Disruption of lipid raft integrity can therefore impair the visual process by disrupting these protein interactions. This can come with huge complications if we don't handle them correctly.

So how do we study these interactions? A variety of biochemical and biophysical techniques are employed. Co-immunoprecipitation allows us to identify proteins that are physically associated with each other. By using an antibody against one protein, we can pull it out of solution along with any proteins it's bound to. Surface plasmon resonance (SPR) is a technique used to measure the binding affinity between two proteins in real-time. Isothermal titration calorimetry (ITC) provides thermodynamic parameters for protein-protein interactions, such as enthalpy and entropy changes, offering insights into the driving forces behind binding. These methods are critical for understanding the dynamics of membrane protein interactions and how they are affected by mutations or drugs. I encourage the use of multiple methods in order to ensure better success.

Furthermore, mutations in genes encoding these interacting proteins are a major cause of inherited retinal diseases. Retinitis pigmentosa (RP), for example, is a group of genetic disorders that lead to progressive photoreceptor cell death and vision loss. Many RP-causing mutations affect the structure or function of rhodopsin or its interacting partners, disrupting their ability to interact properly and triggering a cascade of events that ultimately lead to cell death. Understanding the specific molecular defects caused by these mutations is crucial for developing targeted therapies.

Therefore, we can modulate these interactions using various biotic approaches. Introducing modified lipids can alter the composition and fluidity of the membrane, thereby affecting protein clustering and interactions. Small molecules, such as peptides or peptidomimetics, can be designed to specifically bind to protein interfaces, either enhancing or disrupting their interaction. Gene therapy can be used to deliver corrected versions of genes encoding interacting proteins, restoring their normal function. The possibilities are exciting, but also complex. Consider the development of drugs that stabilize the rhodopsin-transducin interaction in individuals with RP-causing mutations. If a mutation weakens this interaction, a small molecule drug that binds to the interface and strengthens the bond could potentially rescue photoreceptor cell function and slow down disease progression. Alternatively, if a mutation causes excessive activation of rhodopsin, a drug that disrupts the rhodopsin-transducin interaction could prevent overstimulation and cell death. These are just a few examples of how manipulating membrane protein interactions can be used to treat vision impairments.

The ability to manipulate membrane protein interactions opens up exciting possibilities for fine-tuning human vision. But it's essential to proceed with caution, using a multidisciplinary approach that integrates molecular biology, biochemistry, and optical physics. By carefully dissecting the intricate dance of proteins within the photoreceptor cell membrane, we can pave the way for new and innovative therapies that restore and enhance visual function. There is still much work to be done.

3 Chapter 3: Optical Principles in the Eye

The eye, more than just a sensory organ, is an intricate optical instrument. Understanding how light bends and focuses within this biological lens system is paramount to grasping the nuances of vision and, consequently, how impairments arise. Refraction, the bending of light as it passes from one medium to another, is the cornerstone of this process. The cornea, being the eye's primary refractive surface, accounts for a significant portion of the eye's focusing power, directing incoming light rays towards the lens. Following the cornea, the crystalline lens provides further refinement, its ability to change shape, known as accommodation, allowing us to focus on objects at varying distances. This dynamic adjustment is crucial for clear vision, and its loss leads to presbyopia, a common age-related condition. The refractive index of the cornea and lens, along with their curvature, determine the degree to which light is bent. When these parameters are precisely aligned, light converges accurately on the retina, creating a sharp, focused image.

When light waves travel through different mediums, their speed changes, and this change in speed is what causes refraction. In the eye, light first encounters the air, then the cornea, aqueous humor, lens, and finally the vitreous humor before reaching the retina. Each of these interfaces causes a slight bending of light. The cornea, being the most significant contributor to refraction due to its large refractive index difference with air, does most of the work. The lens then fine-tunes the process, especially for near vision. The power of a lens, measured in diopters, indicates its ability to converge or diverge light. A healthy eye possesses the appropriate dioptric power to ensure that light focuses precisely on the retina. However, any mismatch, whether due to corneal shape irregularities, lens abnormalities, or variations in the eye's axial length, leads to refractive errors such as myopia (nearsightedness), hyperopia (farsightedness), and astigmatism. These errors necessitate corrective lenses, either glasses or contact lenses, to compensate for the focusing deficiencies.

Beyond simple refraction, the wave nature of light introduces complexities that further impact image quality. Aberrations, deviations from perfect image formation, arise due to imperfections in the eye's optical components. These imperfections can be broadly categorized into lower-order and higher-order aberrations. Lower-order aberrations, including defocus (myopia and hyperopia) and astigmatism, are relatively simple to correct with conventional eyeglasses or contact lenses. Higher-order aberrations, such as coma, trefoil, and spherical aberration, are more complex and can significantly degrade image quality, even after correcting for lower-order errors. These aberrations arise from subtle irregularities in the shape of the cornea and lens, as well as variations in the refractive index within these structures. Wavefront technology, initially developed for astronomy, has revolutionized our ability to measure and correct higher-order aberrations. By analyzing the wavefront of light exiting the eye, we can create a detailed map of its optical imperfections. This information can then be used to customize corrective lenses or guide laser refractive surgery to reshape the cornea with greater precision.

Considering light as a wave, phenomena like diffraction and interference also play a role, albeit a smaller one, in the eye's optical system. Diffraction, the bending of light around obstacles, can slightly blur the image on the retina, especially at smaller pupil sizes. Interference, the interaction of light waves, can create constructive or destructive patterns, affecting the overall intensity and clarity of the image. These wave optical effects are particularly relevant when considering the impact of corneal irregularities or intraocular lens designs on visual performance. Moreover, the pupil's size plays a crucial role in managing aberrations. A smaller pupil reduces the impact of peripheral aberrations, leading to sharper image quality but also reducing the amount of light entering the eye. Conversely, a larger pupil allows more light to enter, improving vision in low-light conditions, but also increasing the influence of aberrations. This trade-off highlights the complex interplay between the eye's optical components and their effect on overall visual performance. Understanding these optical principles is not only crucial for diagnosing and correcting vision impairments but also for developing innovative therapies that target the underlying causes of these conditions.

Ultimately, a deep understanding of the intricate dance of light within the eye is indispensable. It empowers us to address visual imperfections with targeted precision, moving beyond simple correction towards true visual optimization through novel avenues.

3.1 Refraction and Focusing

The eye, in its remarkable complexity, functions as a sophisticated optical instrument, meticulously crafted to capture and interpret the light that shapes our perception of the world. Refraction, the bending of light as it passes from one medium to another, and focusing, the precise convergence of light rays onto the retina, are the cornerstones of this intricate process, enabling us to see with clarity and detail. Understanding these principles is paramount to grasping how membrane biotics can be leveraged to correct vision impairments arising from irregularities in the eye's optical components. This chapter will delve into the intricacies of refraction and focusing, explaining how these phenomena occur within the eye and how they contribute to both normal and impaired vision.

The journey of light through the eye begins at the cornea, the transparent, dome-shaped outer layer that acts as the primary refractive surface. Due to the significant difference in refractive index between air (approximately 1.0) and the cornea (approximately 1.376), light undergoes substantial bending as it enters this initial interface. This refraction is critical for initiating the focusing process, accounting for approximately 70% of the eye's total refractive power. Consider a ray of light traveling from a distant object; upon encountering the cornea, it is bent inward, starting its trajectory toward the retina. The smoothness and curvature of the cornea are paramount for proper refraction; any irregularities in these characteristics, such as those found in astigmatism, will lead to distorted images. To further illustrate, think of the cornea as a lens maker; it has to have the perfect manufacturing, to achieve a perfect outcome.

Following the cornea, light passes through the aqueous humor, a clear, watery fluid filling the space between the cornea and the lens. The refractive index of the aqueous humor is similar to that of the cornea, minimizing further refraction at this interface. This fluid serves not only as a refractive medium but also provides nutrients to the avascular cornea and lens, maintaining their transparency and functionality. From the aqueous humor, light proceeds to the crystalline lens, a remarkable structure responsible for fine-tuning the focusing process. Unlike the cornea, the lens is flexible and capable of changing its shape, a process known as accommodation. This dynamic adjustment allows the eye to focus on objects at varying distances, ensuring that images are sharp and clear regardless of their proximity. The lens achieves this feat through the action of the ciliary muscles, which surround the lens and control its curvature by altering the tension on the zonular fibers that suspend it. Imagine looking at a close object, such as a book in your hands; the ciliary muscles contract, reducing tension on the zonular fibers, causing the lens to become more convex, increasing its refractive power and bringing the image into focus on the retina. Conversely, when viewing a distant object, the ciliary muscles relax, increasing tension on the zonular fibers, flattening the lens and decreasing its refractive power.

Finally, light traverses the vitreous humor, a clear, gel-like substance filling the space between the lens and the retina. The vitreous humor, with a refractive index similar to that of the aqueous humor, plays a crucial role in maintaining the shape of the eye and supporting the retina. Once through, it finally reaches the retina, the light-sensitive tissue lining the back of the eye. Within the retina, photoreceptor cells, specifically rods and cones, convert light into electrical signals that are then transmitted to the brain via the optic nerve, ultimately resulting in our perception of vision. Accurate focusing is paramount for these signals to be clear and interpretable; if light is not precisely focused on the retina, the resulting image will be blurred.

Aberrations, deviations from perfect focusing, can arise from imperfections in the eye's optical components, such as irregularities in the corneal surface, variations in lens shape or transparency, or even misalignment of the optical axis. These aberrations can be broadly classified into two categories: lower-order aberrations and higher-order aberrations. Lower-order aberrations, including myopia (nearsightedness), hyperopia (farsightedness), and astigmatism, are the most common and can typically be corrected with conventional eyeglasses, contact lenses, or refractive surgery. Myopia occurs when the eye is too long or the cornea is too curved, causing light to focus in front of the retina, resulting in blurry distance vision. Hyperopia, conversely, occurs when the eye is too short or the cornea is too flat, causing light to focus behind the retina, resulting in blurry near vision. Astigmatism arises from an irregularly shaped cornea, causing light to focus at multiple points, resulting in distorted or blurred vision at all distances. On the other hand, higher-order aberrations are more complex and less common and can cause symptoms such as halos, glare, and reduced contrast sensitivity. These aberrations are often more challenging to correct with conventional methods, highlighting the need for advanced approaches such as membrane biotics.

Membrane biotics offers a novel approach to addressing vision impairments by targeting the cellular membranes of the eye's optical components, aiming to improve their structure, function, and transparency. By manipulating the lipid composition and protein content of these membranes, it is possible to enhance their optical properties and minimize aberrations, ultimately leading to improved vision. For instance, in the case of corneal irregularities, membrane biotics could be employed to remodel the corneal surface, smoothing out imperfections and restoring its optimal refractive power. Imagine being able to eliminate astigmatism at its source by directly modifying the corneal cells. Similarly, in the case of lens opacities or cataracts, membrane biotics could be used to enhance lens transparency, improving light transmission to the retina. Furthermore, membrane biotics could be utilized to modulate the flexibility of the lens, improving accommodation and addressing presbyopia, the age-related loss of focusing ability. Thus, it is reasonable to think that the applications of membrane biotics extend far beyond mere vision correction, offering the potential to enhance visual performance and protect against age-related vision decline. As an example, one day, we could prevent the development of cataract, so as to maintain optimal vision for longer.

3.2 Wave Optics and Aberrations

Light, as we know, behaves both as a particle and a wave, and understanding its wave nature is crucial for comprehending how the eye forms images. Geometrical optics, which describes light as rays, is useful for understanding basic refraction and focusing. However, it falls short when explaining phenomena like diffraction and interference, both of which significantly impact the quality of the image formed on the retina. Thinking about light as a wave helps us understand the limits of our vision and provides a foundation for understanding aberrations, the imperfections that degrade image quality. Wave optics gives us a far more accurate picture of the eye's behavior and its limitations, than geometric optics.

The principle of diffraction is very important. It describes how light bends around obstacles and spreads as it passes through an aperture. The pupil of the eye, acting as an aperture, diffracts incoming light. This diffraction is what limits the eye's resolution, even in the absence of any optical aberrations. The smaller the pupil, the more significant the diffraction effects, leading to a blurrier image. This is why our vision is generally sharper in bright light, when the pupil constricts, than in dim light, when it dilates, this is because of the diffraction impact. The Airy disk, a bright central spot surrounded by faint rings, is a result of diffraction. The size of this disk is inversely proportional to the pupil diameter, meaning a smaller pupil results in a larger Airy disk and reduced resolution. The Airy disk represents the smallest point to which a perfect lens can focus a point of light, dictated by the wave nature of light.

Aberrations are deviations from perfect image formation, resulting in blurred or distorted vision. They arise from imperfections in the optical components of the eye, primarily the cornea and the lens. These imperfections cause light rays to deviate from their ideal paths, preventing them from converging perfectly on the retina. Aberrations are generally categorized into lower-order and higher-order aberrations. Lower-order aberrations, which include defocus (myopia and hyperopia) and astigmatism, are the most common and are typically corrected with spectacles or contact lenses. These aberrations are relatively simple to measure and correct, as they primarily involve the overall curvature of the cornea and lens. Higher-order aberrations, on the other hand, are more complex and result from irregularities in the shape of the cornea and lens. These include coma, trefoil, spherical aberration, and others. They are subtler than lower-order aberrations

and were historically more difficult to measure and correct.
Spherical aberration, a particularly important example, occurs when light rays passing through the periphery of a lens (or cornea) are focused at a different point than light rays passing through the center. This results in a blurred image, especially noticeable at night when the pupil dilates and more peripheral light enters the eye. Coma causes objects to appear comet-shaped, while trefoil results in a three-lobed distortion. The effect of higher-order aberrations are generally more noticeable under conditions where the pupil is wider, because more of the aberrant optical surface is being used. If a person has a highly aberrated cornea, then a small pupil will mask much of the aberration.

Measuring aberrations is a crucial step in understanding and correcting visual imperfections. Traditional methods, such as subjective refraction, primarily assess lower-order aberrations. However, advancements in technology have led to the development of wavefront aberrometers, which can measure both lower\item and higher-order aberrations with high precision. Wavefront aberrometry works by projecting a beam of light into the eye and analyzing the reflected wavefront. Any distortions in the wavefront indicate the presence and severity of aberrations. This information is then used to create a "wavefront map" of the eye, which provides a detailed picture of the optical imperfections present. This map is crucial in planning customized vision correction procedures, such as custom LASIK, or in designing specialized contact lenses that can correct for higher-order aberrations.

Correcting for aberrations is a key goal in modern vision correction. Spectacles and traditional contact lenses primarily correct lower-order aberrations, such as myopia, hyperopia, and astigmatism. However, they cannot correct higher-order aberrations. Custom wavefront-guided LASIK surgery uses the information from wavefront aberrometry to reshape the cornea, correcting not only lower-order but also higher-order aberrations. This can result in significantly improved vision quality, particularly in patients with significant higher-order aberrations. Another approach is the use of specialized contact lenses, such as scleral lenses or rigid gas permeable lenses, which can mask corneal irregularities and reduce the impact of aberrations. These lenses create a smooth optical surface, effectively neutralizing the aberrations caused by the underlying cornea. Adaptive optics, a technology originally developed for astronomy, is also being explored for vision correction. Adaptive optics systems use deformable mirrors to compensate for aberrations in real-time, providing the potential for extremely precise and dynamic vision correction.

I must add that the interplay between diffraction and aberrations is complex. Diffraction limits the resolution of the eye, even in the absence of aberrations. Aberrations further degrade image quality by distorting the wavefront of light. In some cases, correcting aberrations can actually worsen vision if diffraction effects are not taken into account. For example, attempting to correct very small amounts of higher-order aberrations may not result in a noticeable improvement in vision, as the improvement may be masked by diffraction. Thus, a comprehensive understanding of both wave optics and aberrations is essential for optimizing vision correction strategies. It is about minimizing the overall impact of these factors in conjunction with each other.

Finally, considering the potential of membrane biotics, imagine being able to selectively modify the lipid composition or protein distribution within the corneal or lens cell membranes to precisely reshape the optical surfaces. This could offer a non-invasive and highly targeted approach to correcting aberrations, going beyond the limitations of current methods. For instance, one could envision delivering specific lipids to regions of the cornea with irregularities, smoothing out the surface and reducing aberrations. Or, perhaps, manipulating membrane proteins to alter the refractive index of specific areas within the lens, creating a customized lens shape that perfectly corrects for the individual's unique aberration profile. This is where the future of vision correction lies: in understanding and manipulating the fundamental building blocks of the eye's optical components at the cellular level. This is a huge advancement in our vision.

4 Chapter 4: Biotic Modulation of Membranes

The intricate dance of lipids and proteins within cellular membranes is not a static tableau; it's a dynamic performance constantly being reshaped by the cellular environment. Our focus here isn't merely on describing that performance, but on learning how to direct it, how to compose a new choreography that enhances visual function. We aim to orchestrate changes in membrane composition and protein expression to address visual impairments. This chapter examines the methods we can use to precisely modulate these membranes for therapeutic benefit.

One powerful approach lies in altering the lipid composition of photoreceptor membranes. The types of lipids present–phospholipids, cholesterol, sphingolipids–directly impact membrane fluidity, permeability, and the organization of membrane proteins. Imagine the cell membrane as a stage; the lipids are the stage floor, affecting how easily the dancers (proteins) can move and interact. Introducing specific lipids, such as polyunsaturated fatty acids (PUFAs), can increase membrane fluidity, potentially improving the diffusion and function of key visual proteins like rhodopsin. Conversely, increasing cholesterol content can stiffen the membrane, which might be beneficial in certain contexts where increased stability is desired, for example, preventing membrane damage in retinal degeneration. This is not a simple additive process; each lipid has a unique shape and charge, and their interactions with each other and with membrane proteins are complex. To achieve specific outcomes, you need to understand the biophysical properties of different lipids and how they influence membrane behavior. We can achieve this through liposome fusion, where pre-formed liposomes containing the desired lipid composition are introduced to the cell. Another approach is through dietary modification or supplementation, which can alter the lipid profile of cell membranes over time. We need to have a solid understanding on the type of lipid and the best possible way to introduce it.

Beyond just changing the lipid environment, we can directly manipulate the protein content of the membrane. Protein insertion techniques allow us to introduce new proteins or modify existing ones to enhance visual function. This can be achieved through viral vectors, which deliver genes encoding the desired proteins directly into the cell. Imagine delivering blueprints to the cell's construction crew, instructing them to build new components for our membrane stage. For example, in cases of inherited retinal degenerations caused by mutations in specific membrane proteins, we could use gene therapy to deliver a functional copy of the gene, leading to the production of the correct protein and restoration of visual function. Another approach is to use protein transduction domains (PTDs), short amino acid sequences that can facilitate the direct uptake of proteins across the cell membrane. This can be useful for delivering therapeutic proteins or enzymes that can modify existing membrane proteins. However, protein insertion is not without its challenges. Ensuring proper protein folding, trafficking to the correct location within the membrane, and maintaining long-term stability are all critical considerations. Moreover, immune responses to the introduced proteins need to be carefully managed.

The true power of membrane biotics lies in its potential for targeted drug delivery. Delivering therapeutic agents directly to the photoreceptor membranes minimizes off-target effects and maximizes the drug concentration at the site of action. We can use nanotechnology to create drug-loaded nanoparticles that are specifically designed to bind to the surface of photoreceptor cells. Imagine these nanoparticles as tiny guided missiles, delivering their payload precisely where it's needed. The surface of these nanoparticles can be decorated with antibodies or ligands that recognize specific proteins on the photoreceptor cell surface. Once the nanoparticles bind to the cells, they can be internalized through endocytosis, releasing their drug cargo inside the cell. Another approach is to use light-activated drugs, also known as photopharmaceuticals. These drugs are inactive until they are exposed to light of a specific wavelength. By focusing light on the photoreceptor membranes, we can selectively activate the drug in the desired location, minimizing systemic exposure and side effects. For example, light-activated drugs could be used to inhibit specific enzymes involved in retinal degeneration or to promote the regeneration of photoreceptor cells.

However, modulating membranes is not without its potential pitfalls. Off-target effects, unintended consequences, and long-term stability are all important considerations. Imagine tinkering with a complex machine; a small adjustment in one area can have unforeseen effects elsewhere. Therefore, thorough preclinical testing is essential to evaluate the safety and efficacy of membrane biotics interventions before they can be translated to human patients. This includes in vitro studies to assess the effects of membrane modulation on cellular function and animal studies to evaluate the effects on visual function and retinal health. Only through rigorous scientific investigation can we unlock the full potential of membrane biotics for treating vision impairments.

Think of the eye as a finely tuned instrument. The photoreceptor membranes are the strings of this instrument, and their composition and function directly determine the quality of the music - our vision. By carefully modulating these membranes, we can fine-tune the instrument, correcting imperfections and restoring harmony. This approach represents a paradigm shift in vision care, moving beyond traditional corrective lenses and surgery to address the underlying molecular causes of visual dysfunction. The ability to precisely manipulate the structure and function of photoreceptor membranes opens up new avenues for treating a wide range of vision impairments, from common refractive errors to devastating retinal degenerations. It requires a deep understanding of membrane biology, optical physics, and pharmacology, but the potential rewards are immense. Through careful research and development, membrane biotics has the potential to transform the lives of millions of people suffering from vision loss, enabling them to see the world more clearly and experience the richness of life.

4.1 Lipid Composition Alterations

Lipid composition is paramount for proper membrane function, directly impacting fluidity, protein interactions, and overall cellular signaling. Alterations in lipid composition, whether through dietary influence, genetic predisposition, or disease processes, can profoundly disrupt the delicate balance required for optimal vision. In the context of photoreceptor membranes, with their high content of polyunsaturated fatty acids (PUFAs), these alterations are particularly critical, leading to a cascade of events that can ultimately impair visual acuity and contribute to retinal degeneration. These tiny, almost invisible, molecules are much more than simply fats.

Understanding the specific lipid changes associated with various vision impairments is crucial for developing targeted interventions. For instance, age-related macular degeneration (AMD), a leading cause of vision loss, is often associated with decreased levels of long-chain PUFAs like docosahexaenoic acid (DHA) in photoreceptor membranes. DHA, an omega-3 fatty acid, is crucial for maintaining membrane fluidity and the proper functioning of rhodopsin, the light-sensitive protein in rods. A deficiency in DHA can lead to increased oxidative stress, inflammation, and ultimately, photoreceptor cell death. Conversely, an excess of saturated fatty acids can stiffen the membrane, hindering protein mobility and signal transduction. Similarly, conditions like retinitis pigmentosa (RP), a group of inherited retinal diseases, can be linked to specific mutations affecting lipid metabolism, leading to abnormal lipid accumulation or depletion within photoreceptor cells. The specific lipids that are impacted can vary considerably, resulting in an array of disease phenotypes. Analyzing the lipid profiles of affected individuals is a critical step in understanding the underlying pathology and informing potential therapeutic strategies. We also have to remember that lipid composition is not a static property; it is constantly being remodeled and adjusted by cellular enzymes. Manipulating lipid composition represents a promising avenue for therapeutic intervention in vision impairments. One approach involves dietary supplementation with specific lipids, particularly omega-3 fatty acids. Clinical trials have shown that increasing DHA intake can slow the progression of AMD in some individuals. However, the efficacy of dietary supplementation can vary depending on individual factors such as genetics, age, and the severity of the condition. Another strategy focuses on directly modulating the activity of enzymes involved in lipid metabolism. For example, inhibiting enzymes that promote the synthesis of saturated fatty acids could help to restore a more favorable lipid balance in photoreceptor membranes. This approach requires careful consideration of potential off-target effects and the need for tissue-specific targeting to avoid disrupting lipid metabolism in other parts of the body. Furthermore, the delivery of specific lipids directly to photoreceptor cells using nanoparticles or other targeted delivery systems holds great promise for enhancing therapeutic efficacy. This would allow for precise control over lipid composition within the target tissue, minimizing systemic side effects.

Protein-lipid interactions are fundamental to membrane function, and alterations in lipid composition can significantly impact these interactions. Many membrane proteins, including rhodopsin, rely on specific lipid domains for proper localization, folding, and activity. Changes in lipid saturation, charge, or headgroup size can disrupt these interactions, leading to protein mislocalization or dysfunction. For instance, cholesterol, a major component of cell membranes, plays a critical role in organizing lipids into distinct domains known as lipid rafts. These rafts serve as platforms for the assembly of signaling complexes and can influence the activity of membrane proteins. Disruptions in cholesterol homeostasis can alter the structure and function of lipid rafts, leading to impaired signal transduction in photoreceptor cells. Moreover, specific lipids can directly bind to membrane proteins, modulating their activity. Phosphatidylinositol phosphates (PIPs), for example, are signaling lipids that bind to a variety of proteins, regulating their localization and function. Changes in PIP levels or distribution can disrupt these interactions, leading to aberrant signaling and cellular dysfunction. Comprehending how different lipid species interact with membrane proteins is crucial for developing targeted therapies that can restore proper protein function in the context of altered lipid composition.

Beyond photoreceptor cells, the lipid composition of other retinal cells, such as retinal pigment epithelium (RPE) cells, also plays a critical role in vision. RPE cells form a monolayer that supports and nourishes photoreceptor cells, and their lipid composition is essential for maintaining their phagocytic function, which is responsible for clearing shed photoreceptor outer segments. Alterations in RPE lipid composition can impair phagocytosis, leading to the accumulation of debris and the formation of drusen, a hallmark of AMD. Moreover, RPE cells are responsible for transporting lipids and other nutrients to photoreceptor cells, and changes in their lipid transport mechanisms can disrupt photoreceptor metabolism. Therefore, targeting lipid composition in RPE cells represents another promising avenue for treating vision impairments. This complex interplay highlights the importance of considering the entire retinal microenvironment when developing membrane biotics therapies.

Future research should focus on identifying novel lipid targets and developing more sophisticated strategies for modulating lipid composition in a tissue-specific and cell-type-specific manner. The use of advanced imaging techniques, such as mass spectrometry imaging, can provide detailed information about the spatial distribution of lipids within the retina, allowing for a better understanding of how lipid alterations contribute to disease pathogenesis. Furthermore, the development of new lipid-based nanoparticles and other delivery systems can enable the precise targeting of therapeutic lipids to specific retinal cells. These advances will pave the way for more effective and personalized membrane biotics therapies for a wide range of vision impairments. Ultimately, the ability to fine-tune lipid composition in the retina holds immense potential for preventing and treating vision loss, improving the quality of life for millions of people worldwide.

4.2 Protein Insertion Techniques

Protein insertion is not just about getting a protein into a membrane; it's about precision, control, and ultimately, therapeutic effect. For beginners entering the field of membrane biotics, grasping the nuances of these techniques is paramount, especially when targeting photoreceptor cells. Here, we will explore various methods, focusing on their practical application and potential challenges in the context of vision impairment.

One of the more established methods is the use of liposomes. These spherical vesicles, composed of lipid bilayers, can encapsulate proteins and fuse with the target cell membrane, delivering their cargo directly. Think of them as tiny, biocompatible bubbles designed for a specific mission. The beauty of liposomes lies in their versatility; we can tailor their lipid composition to enhance fusion with specific cell types or even modify their surface with targeting ligands to ensure they reach the intended destination. The insertion of proteins into liposomes themselves can be achieved through various methods, including sonication or extrusion, forcing the protein into the lipid bilayer during liposome formation. We must consider the protein's structure and orientation; ensuring the protein is correctly oriented within the membrane after liposome fusion is crucial for its function. It's a bit like ensuring a key is inserted into a lock the right way up.

Another technique that warrants attention is the use of viral vectors. Adeno-associated viruses (AAVs), in particular, have gained traction due to their low immunogenicity and ability to transduce a wide range of cell types, including retinal cells. The process involves packaging the gene encoding the desired protein into the AAV vector, which then infects the target cells and delivers the genetic material. The cell then uses its own machinery to synthesize the protein and insert it into the membrane. While AAVs are effective, careful consideration must be given to the serotype used, as different serotypes exhibit varying tropism for different cell types within the retina. Moreover, controlling the expression levels of the inserted protein is crucial to avoid potential toxic effects. It's a balancing act, ensuring enough protein is produced to exert a therapeutic effect without overwhelming the cell. The choice of promoter also plays a vital role in regulating gene expression; using a cell-specific promoter can further enhance targeting and minimize off-target effects.

Cell-penetrating peptides (CPPs) offer a non-viral approach to protein delivery. These short amino acid sequences can translocate across cell membranes, carrying with them attached proteins or other cargo. One of the advantages of CPPs is their relative simplicity and ease of use. We can chemically conjugate CPPs to the protein of interest and administer them directly to the cells or tissues. However, CPP-mediated delivery is often less efficient than viral transduction, and the mechanism of translocation is not fully understood. Some CPPs may enter cells via endocytosis, resulting in the protein being trapped within vesicles rather than inserted into the membrane. Optimizing the CPP sequence and conjugation strategy can improve delivery efficiency, but it is a technique that requires careful optimization and validation. Consider the charge and hydrophobicity of the CPP, as these factors can influence its ability to interact with the cell membrane.

Beyond these established methods, researchers are exploring more advanced techniques, such as the use of nanocarriers. These tiny particles, made of materials like gold, silica, or polymers, can be engineered to deliver proteins directly to the cell membrane. Nanocarriers offer the potential for highly targeted delivery and controlled release of the protein. We can modify their surface with targeting ligands to enhance binding to specific receptors on the cell surface, and they can be designed to release their cargo in response to specific stimuli, such as pH or light. While nanocarriers hold great promise, their development is still in its early stages. Challenges remain in terms of biocompatibility, toxicity, and scalability. It's a frontier of innovation, pushing the boundaries of what's possible in protein delivery. Think of nanocarriers as miniature submarines, navigating the complex cellular landscape to deliver their precious cargo.

When it comes to applying these techniques to address vision impairments, we must carefully consider the specific needs of the target cells. Photoreceptor cells, for example, are highly specialized cells with unique membrane properties. Their outer segment membranes, where phototransduction occurs, are constantly being renewed, making them a dynamic and challenging target for protein insertion. Furthermore, the retina is a delicate tissue, and any protein insertion strategy must be minimally invasive to avoid causing further damage. The choice of delivery method, the dose of protein administered, and the route of administration must all be carefully optimized to ensure safety and efficacy. It's a delicate dance, balancing the need to deliver therapeutic proteins with the imperative to protect the integrity of the eye.

Another layer of complexity arises from the inherent variability between individuals. Genetic differences, environmental factors, and the specific nature of the vision impairment can all influence the effectiveness of protein insertion strategies. Personalized medicine, tailoring treatment to the individual patient, is likely to play an increasingly important role in membrane biotics. This may involve using patient-derived cells or tissues to test different protein insertion strategies in vitro before administering them in vivo. It could also involve developing biomarkers to predict which patients are most likely to respond to a particular treatment. The future of membrane biotics lies in personalized approaches, recognizing that each patient is unique and requires a tailored solution. This is not a one-size-fits-all field; it demands a nuanced and individualized approach.

Finally, it's important to acknowledge the ethical considerations surrounding protein insertion techniques. While these techniques hold great promise for treating vision impairments, they also raise questions about safety, accessibility, and the potential for unintended consequences. Ensuring that these technologies are developed and used responsibly is crucial for maintaining public trust and maximizing their benefit to society. We must engage in open and honest discussions about the ethical implications of membrane biotics and ensure that patient safety and well-being are always paramount. It's not just about what we can do, but what we should do, guided by ethical principles and a commitment to improving the lives of those with vision impairments.

4.3 Targeted Drug Delivery

Targeted drug delivery represents a pivotal aspect of biotic modulation of membranes, especially when seeking to fine-tune human vision impairments with optical mechanisms. Conventional drug administration often suffers from limited bioavailability, off-target effects, and insufficient concentrations reaching the affected retinal cells or choroidal vasculature. This is particularly problematic when addressing delicate and highly localized pathologies within the eye, emphasizing the necessity for strategies that ensure precise delivery to the intended cellular targets. To realize this vision, researchers are exploring various approaches that rely on nanotechnology, molecular targeting, and advanced drug formulation techniques.

The limitations of systemic drug delivery in ophthalmology stem from the blood-retinal barrier (BRB), a formidable physiological barrier that restricts the passage of many substances from the bloodstream into the retina. This barrier, formed by tight junctions between endothelial cells of retinal capillaries and retinal pigment epithelial cells, effectively prevents many therapeutic agents from reaching their intended targets. This challenge can be somewhat circumvented by local delivery methods, such as intravitreal injections, but these invasive procedures carry their own risks, including retinal detachment, endophthalmitis, and cataracts. Furthermore, while intravitreal injections allow for a higher concentration of drug to reach the retina initially, the drug is often cleared rapidly, requiring frequent injections which can be burdensome for patients and increase the risk of complications. Targeted drug delivery systems aim to overcome these limitations by providing a means to transport therapeutic agents specifically to diseased cells or tissues, minimizing off-target effects and maximizing therapeutic efficacy.

Nanoparticles have emerged as a promising platform for targeted drug delivery in ophthalmology. These particles, typically ranging in size from 1 to 1000 nanometers, can be engineered to encapsulate therapeutic agents and functionalized with targeting ligands that recognize specific receptors or antigens expressed on the surface of target cells. For example, nanoparticles coated with antibodies against vascular endothelial growth factor (VEGF) receptors can be used to selectively deliver anti-VEGF drugs to retinal endothelial cells in patients with neovascular age-related macular degeneration (AMD). These VEGF inhibitors reduce blood vessels growth. Different materials, such as liposomes, polymeric nanoparticles, and inorganic nanoparticles, can be used to fabricate these drug carriers. Liposomes, composed of lipid bilayers, are biocompatible and can encapsulate both hydrophilic and hydrophobic drugs. Polymeric nanoparticles, made from biodegradable polymers such as polylactic-co-glycolic acid (PLGA), offer controlled drug release and can be easily modified with targeting ligands. Inorganic nanoparticles, such as gold nanoparticles and quantum dots, possess unique optical properties that can be exploited for imaging and photothermal therapy. By attaching specific peptides to the surface of these nanoparticles, researchers can further enhance their targeting specificity and improve their ability to cross the BRB.

Beyond nanoparticles, other approaches to targeted drug delivery include the use of aptamers and cell-penetrating peptides (CPPs). Aptamers are short, single-stranded DNA or RNA molecules that can bind to specific target molecules with high affinity and specificity. They can be conjugated to therapeutic agents or nanoparticles to guide them to the desired location within the eye. CPPs, on the other hand, are short amino acid sequences that can facilitate the translocation of cargo molecules across cell membranes. By attaching CPPs to therapeutic agents, it is possible to enhance their cellular uptake and improve their therapeutic efficacy. The use of aptamers specific to cell-surface markers overexpressed in diseased retinal cells, such as those found in proliferative diabetic retinopathy, allows for selective delivery of therapeutic payloads to these cells while sparing healthy tissue. Similarly, CPPs can be employed to enhance the penetration of drugs across the BRB, thereby increasing the drug concentration in the retina.

Effective targeted drug delivery requires not only precise targeting but also controlled drug release. Premature release of the drug can lead to off-target effects and reduced therapeutic efficacy, while delayed release may result in insufficient drug concentrations at the target site. To address this challenge, researchers are developing stimuli-responsive drug delivery systems that release their payload only upon encountering a specific trigger in the target environment. These triggers can include changes in pH, temperature, light, or the presence of specific enzymes. For instance, pH-sensitive nanoparticles can be designed to release their drug payload in the acidic microenvironment of tumors or inflamed tissues. Light-activated drug delivery systems can be used to release drugs only in areas exposed to a specific wavelength of light, allowing for precise spatial control over drug release. Temperature-sensitive liposomes can release their contents upon reaching a certain temperature, allowing for targeted delivery to hyperthermic tissues. These strategies enable a more tailored and controlled release of therapeutic agents, maximizing their effectiveness while minimizing unwanted side effects.

The development of targeted drug delivery systems for vision impairments is a rapidly evolving field, with significant potential to improve the treatment of a wide range of ocular diseases. However, several challenges remain. One major challenge is the complex nature of the eye and the presence of multiple barriers that drugs must overcome to reach their intended targets. The BRB, in particular, poses a significant obstacle to drug delivery, and researchers are actively exploring strategies to enhance the penetration of drugs across this barrier. Another challenge is the potential for off-target effects, even with targeted drug delivery systems. It is crucial to carefully evaluate the safety and efficacy of these systems in preclinical and clinical studies to ensure that they do not cause unintended harm. Furthermore, the cost of developing and manufacturing these advanced drug delivery systems can be high, which may limit their accessibility to patients. Despite these challenges, the potential benefits of targeted drug delivery for vision impairments are immense, and continued research in this area is essential to improve the lives of patients with blinding eye diseases.

5 Chapter 5: Fine-Tuning for Vision Correction

The human eye, a marvel of biological engineering, is susceptible to a range of refractive errors, each stemming from imperfections in the eye's ability to focus light precisely on the retina; therefore, the application of membrane biotics holds promise for targeted interventions, offering the potential for refined vision correction, and in cases such as myopia, which typically arises from an elongated eyeball or an excessively curved cornea, light converges in front of the retina, resulting in blurred distance vision, we can consider modulation of scleral fibroblast activity through targeted delivery of specific lipids; these lipids could influence the extracellular matrix composition, potentially remodeling the sclera to reduce axial length and shift the focal point back onto the retina. Furthermore, we might explore enhancing the elasticity of the choroid, the vascular layer beneath the retina, to facilitate scleral remodeling during natural eye movements; this could involve introducing specific proteins or peptides that promote choroidal blood flow and nutrient delivery, improving tissue health and responsiveness to biomechanical cues. Similarly, hyperopia, or farsightedness, occurs when the eyeball is too short or the cornea is too flat, causing light to focus behind the retina; in this scenario, membrane biotics could be employed to stiffen the sclera, preventing further elongation and potentially improving the eye's natural focusing ability, besides, we could investigate promoting the production of specific extracellular matrix components that enhance scleral rigidity and resistance to deformation. Also, targeted drug delivery could be utilized to deliver growth factors that stimulate the proliferation and differentiation of corneal endothelial cells; this could thicken the cornea, increasing its refractive power and shifting the focal point forward onto the retina.

Moreover, astigmatism, characterized by an irregularly shaped cornea or lens, causes light to focus unevenly on the retina, resulting in distorted or blurred vision at all distances; in this case, a customized approach is crucial, and membrane biotics could involve precisely modulating the lipid composition of specific regions of the cornea, guided by sophisticated corneal topography mapping. This targeted manipulation could reshape the corneal surface, correcting the refractive irregularities that cause astigmatism; imagine a nanosurgical technique, where liposomes containing specific lipid components are delivered to the corneal stroma, guided by real-time imaging feedback. These liposomes would fuse with the stromal cell membranes, altering the local lipid environment and influencing the organization of collagen fibrils, gradually reshaping the corneal surface and correcting the astigmatism. In addition to corneal reshaping, we could explore the potential of modulating the lens capsule's elasticity; the lens capsule is a thin, transparent membrane that surrounds the crystalline lens. By altering its composition, we might be able to induce subtle changes in lens shape, further refining the correction of astigmatism.

Presbyopia, the age-related loss of accommodation, presents a unique challenge; as the crystalline lens loses its elasticity, it becomes difficult to focus on near objects; instead of addressing refractive errors, the focus shifts to restoring lens flexibility; therefore, membrane biotics could involve introducing specific enzymes that break down cross-linked proteins within the lens, restoring its ability to change shape; imagine injecting a biocompatible hydrogel containing these enzymes into the lens capsule; the enzymes would selectively degrade the cross-linked proteins, softening the lens and improving its ability to accommodate. Another approach could involve stimulating the production of chaperone proteins within the lens; chaperone proteins assist in the proper folding and assembly of other proteins, preventing aggregation and maintaining lens transparency; by enhancing the expression of these chaperone proteins, we could improve the overall health and function of the lens, delaying the onset of presbyopia. This could involve delivering specific growth factors or signaling molecules that activate intracellular signaling pathways, promoting chaperone protein synthesis.

Furthermore, color vision deficiencies, often stemming from abnormalities in the cone photoreceptors of the retina, pose a different set of challenges; while complete cures may be elusive, membrane biotics could potentially enhance the function of existing cone cells or even introduce new light-sensitive pigments; in this case, gene therapy is critical, where we could deliver genes encoding missing or dysfunctional opsins, the light-sensitive proteins in cone cells. These genes would be delivered via viral vectors or nanoparticles, targeting specific cone cell populations. Once inside the cells, the genes would be expressed, producing functional opsins and restoring partial color vision. Alternatively, we could explore the potential of using light-sensitive molecules that bind to cell membranes and respond to specific wavelengths of light. These molecules could act as artificial photoreceptors, supplementing the function of existing cone cells and expanding the range of colors that can be perceived.

Retinal degeneration, a devastating group of diseases that progressively damage the retina, requires a multifaceted approach; membrane biotics could play a role in protecting remaining photoreceptors, promoting their survival, and even stimulating the regeneration of damaged tissue; this could involve delivering neurotrophic factors, proteins that support the survival and growth of neurons, directly to the retina; these factors would protect photoreceptors from apoptosis, or programmed cell death, preventing further vision loss, besides, we could explore the potential of using stem cells to replace damaged photoreceptors. Stem cells could be differentiated into photoreceptor-like cells and transplanted into the retina, where they would integrate into the existing neural circuitry and restore visual function. Furthermore, the use of exosomes could play a crucial role; exosomes are small vesicles secreted by cells that contain a variety of molecules, including proteins, lipids, and nucleic acids; these exosomes can be used to deliver therapeutic agents directly to the retina, protecting photoreceptors, promoting their survival, and stimulating the regeneration of damaged tissue. These therapeutic agents can then protect photoreceptors from apoptosis, preventing further vision loss. In this regard, the chapter offers practical and actionable insights for those embarking on membrane biotics for vision correction, providing a glimpse into the potential of this exciting field.

5.1 Myopia and Hyperopia

Myopia, or nearsightedness, and hyperopia, or farsightedness, represent the most common refractive errors affecting global vision. They arise from a mismatch between the eye's optical power and its axial length, preventing light from focusing correctly on the retina. In myopia, the eye is too long relative to its refractive power, causing light to focus in front of the retina, resulting in blurry distance vision. Conversely, in hyperopia, the eye is too short, or the refractive power is too weak, leading to light focusing behind the retina, impacting near vision primarily but often also affecting distance vision to some degree, particularly with increasing age. Traditional correction methods, such as glasses and contact lenses, simply redirect light to compensate for these errors, but do not address the underlying structural or functional issues within the eye itself. Refractive surgeries offer a more permanent solution by reshaping the cornea, but they are invasive and carry potential risks. This is where the potential of membrane biotics comes into play, offering a novel approach to address the root causes of these conditions at the cellular level.

The development of myopia and hyperopia is a complex process influenced by a combination of genetic and environmental factors. Studies have shown a strong hereditary component, suggesting that individuals with parents who have refractive errors are more likely to develop them themselves. However, genetics alone cannot fully explain the increasing prevalence of myopia, especially in East Asian countries where rates have skyrocketed in recent decades. Environmental factors, such as increased near work activities (reading, screen time) and decreased time spent outdoors, are believed to play a significant role. These activities can place prolonged strain on the eye's focusing mechanisms, potentially leading to structural changes that contribute to myopia development. Research suggests that exposure to natural light triggers the release of dopamine in the retina, which inhibits eye elongation, offering a protective effect against myopia. For hyperopia, genetic predisposition is considered a primary factor, although the precise mechanisms are not fully understood. It is important to note that in some cases, hyperopia in children can be a normal developmental stage that resolves as the eye grows and matures. The intricate interplay between genetics and environmental factors highlights the need for multi-faceted strategies to prevent and manage these refractive errors.

Membrane biotics offers exciting possibilities for addressing myopia and hyperopia by targeting the cellular mechanisms involved in eye growth and refractive development. One potential approach involves modulating the lipid composition of scleral fibroblasts, the cells responsible for producing the sclera, the tough outer coat of the eye. Altering the lipid profile could influence scleral rigidity and remodeling, thereby controlling eye elongation in myopia or promoting appropriate growth in hyperopia. Specifically, incorporating lipids that enhance scleral crosslinking could increase its stiffness, resisting excessive elongation in myopic eyes. Another avenue of exploration is the manipulation of membrane protein interactions within the retina. Retinal pigment epithelium (RPE) cells, which play a critical role in maintaining photoreceptor health and regulating eye growth, express a variety of membrane proteins involved in signaling pathways that influence scleral remodeling. By modulating the activity of these proteins through targeted drug delivery or gene therapy approaches, we could potentially influence eye growth and refractive state. For example, enhancing the expression of proteins that promote scleral collagen synthesis could help to correct hyperopia by promoting eye growth. These strategies aim to directly influence the biological processes that contribute to refractive error development, offering a more targeted and potentially more effective approach than traditional methods.

Targeted drug delivery is a crucial aspect of membrane biotics, ensuring that therapeutic agents reach the intended cells and membranes within the eye. Nanoparticles, liposomes, and other biocompatible carriers can be engineered to deliver drugs specifically to scleral fibroblasts or RPE cells, minimizing off-target effects and maximizing therapeutic efficacy. These carriers can be functionalized with targeting ligands that bind to specific receptors on the surface of these cells, enhancing their uptake and localization. Moreover, the release of the drug from the carrier can be controlled by external stimuli, such as light or ultrasound, allowing for precise temporal and spatial control over drug delivery. For instance, light-activated nanoparticles could be injected into the sclera and then activated by shining a specific wavelength of light onto the eye, triggering the release of drugs that modulate scleral remodeling. This level of precision would minimize systemic exposure and reduce the risk of side effects. The development of advanced drug delivery systems is essential for the successful translation of membrane biotics into clinical practice.

In vivo studies using animal models are crucial for evaluating the safety and efficacy of membrane biotics interventions for myopia and hyperopia. Animal models, such as guinea pigs and chickens, can be induced to develop myopia or hyperopia through specific environmental manipulations, such as altering visual input or rearing conditions. These models allow researchers to assess the effects of membrane-modulating therapies on eye growth, refractive error, and retinal function. Histological analysis of the sclera and retina can provide insights into the cellular and molecular mechanisms underlying the observed effects. Furthermore, electrophysiological measurements can be used to assess the impact of these therapies on visual function and retinal signaling. These studies provide valuable preclinical data that are essential for guiding the development of clinical trials in humans.

Ultimately, the goal of membrane biotics is to develop safe and effective therapies for preventing and treating myopia and hyperopia in humans. Clinical trials are necessary to evaluate the safety, efficacy, and long-term effects of these interventions. These trials would involve carefully selected participants with myopia or hyperopia who would receive either the membrane-modulating therapy or a placebo control. Changes in refractive error, eye growth, and visual function would be carefully monitored over time. Moreover, safety assessments would be conducted to identify and manage any potential side effects. Ethical considerations are paramount in clinical trial design, ensuring that participants are fully informed about the risks and benefits of the study and that their privacy and well-being are protected. The successful completion of clinical trials is essential for bringing membrane biotics from the laboratory to the clinic and making these potentially transformative therapies available to patients.

5.2 Astigmatism Correction

Astigmatism arises from an irregularly shaped cornea, or less commonly, the lens, causing light to focus unevenly onto the retina. Instead of a single focal point, light rays converge at multiple points, resulting in blurred or distorted vision at all distances. The degree and orientation of this distortion vary from person to person, making personalized correction paramount. Traditional methods such as eyeglasses and contact lenses work by adding a corrective cylindrical power to compensate for the eye's irregular shape, refracting light in a way that creates a single, clearer focal point on the retina. While effective, these methods are essentially external prosthetics, requiring constant reliance and, in the case of contact lenses, meticulous hygiene. Refractive surgeries, like LASIK and PRK, reshape the cornea to correct astigmatism, offering a more permanent solution but involve invasive procedures with potential risks and limitations, particularly for high degrees of astigmatism or individuals with thin corneas. Understanding the nuanced nature of astigmatism and the limitations of current corrective measures is crucial before exploring the potential of membrane biotics.

Our approach focuses on modulating the cellular membranes within the cornea to reshape its surface at a microscopic level, aiming to improve the eye's natural focusing ability. Unlike external corrections or macro-level reshaping, membrane biotics offers the potential for a more subtle, biocompatible, and potentially long-lasting solution. We propose altering the lipid composition of corneal cell membranes to influence cell shape and arrangement. By introducing specific lipids or modulating the existing lipid profile, we can influence membrane fluidity and curvature, effectively guiding the corneal cells to reorganize into a more regular shape. This involves targeted delivery of lipid precursors or enzymes that catalyze lipid modification directly to the affected corneal areas, utilizing nanotechnology-based carriers to ensure precise and efficient delivery. Imagine, instead of grinding away corneal tissue, we are gently coaxing it into a more functional form.

The key to success lies in precisely targeting the areas of the cornea responsible for the astigmatism. Corneal topography, a detailed mapping of the cornea's surface, provides the necessary data to identify these specific zones. With this information, we can design targeted delivery systems that release the membrane-modulating agents only in the desired locations. These systems could employ various mechanisms, such as light-activated release, where a focused beam of light triggers the release of the therapeutic agent, or receptor-mediated delivery, where the carrier molecule binds to specific receptors on the corneal cells in the target area. Moreover, we are exploring the use of micro-needles for precise injection of these agents into the corneal stroma, bypassing the epithelial barrier and ensuring deeper penetration and more localized effect.

Protein insertion techniques offer another avenue for correcting astigmatism at the membrane level. By introducing or modifying membrane proteins within corneal cells, we can directly influence cell shape and interactions. For instance, we could introduce proteins that promote cell adhesion in specific regions, encouraging the corneal cells to align in a more uniform manner. Alternatively, we could modify existing membrane proteins to alter their interactions with the cytoskeleton, the internal scaffolding of the cell, thus indirectly influencing cell shape and corneal curvature. This approach requires a deep understanding of the specific proteins involved in corneal cell structure and function, as well as the development of efficient methods for protein delivery and integration into the cell membrane.

Before clinical application, extensive in vitro and in vivo studies are crucial. In vitro studies using cultured corneal cells allow us to investigate the effects of different membrane-modulating agents on cell shape, adhesion, and refractive properties. Animal models of astigmatism, created through surgical or genetic manipulation, provide a platform for testing the efficacy and safety of our membrane biotics approach in a living organism. These studies involve detailed measurements of corneal topography, optical coherence tomography (OCT) imaging to assess corneal structure, and electroretinography (ERG) to evaluate retinal function. The goal is to demonstrate that our approach can effectively reduce astigmatism without causing any adverse effects on corneal health or visual function.

While promising, membrane biotics for astigmatism correction faces several challenges. Precisely controlling the cellular response to membrane modulation is crucial, as over\item or under-correction could lead to undesirable visual outcomes. Ensuring long-term stability of the corrected corneal shape is also essential, as the cornea is a dynamic tissue that undergoes constant remodeling. Furthermore, addressing potential immune responses to the introduced lipids or proteins is vital for preventing inflammation and rejection. Long-term studies are needed to assess the durability and safety of the treatment. But, I believe that the potential benefits of this approach – a non-invasive, biocompatible, and personalized solution for astigmatism correction – make it a worthwhile pursuit.

Imagine a future where astigmatism is not corrected with external devices or invasive surgery, but rather with a gentle nudge to the eye's own cells. This is the promise of membrane biotics, and it is a vision worth striving for.

5.3 Presbyopia Solutions

The gradual loss of accommodative amplitude, the ability to focus on near objects, is the hallmark of presbyopia. It affects everyone, typically becoming noticeable in the early to mid-40s, as the crystalline lens within the eye loses its elasticity and the ciliary muscle, responsible for lens shape adjustment, weakens. While conventional treatments like reading glasses, multifocal lenses, and refractive surgery offer solutions, membrane biotics presents a novel approach by targeting the underlying biological mechanisms of this aging process. Instead of merely correcting the optical error, we aim to restore or enhance the lens's flexibility and accommodative power.

The crystalline lens capsule, composed primarily of collagen, is the outer envelope surrounding the lens fibers. Its elasticity plays a vital role in accommodation. As we age, cross-linking within the collagen fibers increases, reducing the capsule's ability to deform and transmit the force exerted by the ciliary muscle. Imagine a rubber band; with age, it becomes stiff and loses its stretchiness. One potential avenue for membrane biotics involves the enzymatic cleavage of these excessive cross-links. Delivering specific enzymes, encapsulated in biocompatible nanocarriers, directly to the lens capsule could selectively break down the cross-links, restoring some of its original elasticity. The challenge lies in ensuring precise targeting and controlling the enzymatic activity to avoid over-correction or damage to surrounding tissues. Furthermore, the delivery mechanism needs to overcome the blood-aqueous barrier, which protects the eye from foreign substances.

Lens epithelial cells (LECs), located beneath the anterior lens capsule, are responsible for producing lens proteins and maintaining lens homeostasis. With age, LEC function declines, contributing to lens stiffening and cataract formation. Manipulating the LEC membrane composition could enhance their functional capacity. Incorporating specific lipids, like omega-3 fatty acids, into the LEC membranes could improve their fluidity and resilience. This, in turn, might enhance protein production and reduce the accumulation of damaged proteins that contribute to lens opacity. Consider this as providing the LECs with the essential building blocks they need to function optimally. The key is to deliver these lipids in a form that can be efficiently incorporated into the cell membranes, bypassing any metabolic barriers.

Another exciting possibility involves manipulating the lens fiber cells themselves. These highly specialized cells, packed with crystallin proteins, contribute significantly to the lens's refractive properties. As these proteins age, they can aggregate and become less soluble, leading to light scattering and reduced lens transparency. Promoting the proper folding and preventing the aggregation of crystallin proteins could maintain lens clarity and flexibility. Chaperone proteins, responsible for assisting in protein folding, could be delivered to the lens fiber cells to enhance this process. Alternatively, small molecules that bind to crystallin proteins and prevent their aggregation could be introduced. Imagine these molecules as tiny "molecular chaperones" that guide the crystallin proteins into their correct shape, preventing them from clumping together. The success of this approach hinges on identifying the specific crystallin proteins that are most prone to aggregation in presbyopia and designing targeted interventions.

The ciliary muscle, responsible for controlling the lens's shape during accommodation, weakens with age. Enhancing ciliary muscle function could partially compensate for the lens's reduced elasticity. Membrane biotics could potentially target the neuromuscular junctions of the ciliary muscle, improving the efficiency of nerve impulse transmission. Delivering growth factors or neurotransmitters that promote muscle fiber regeneration or enhance acetylcholine release could stimulate ciliary muscle activity. Think of it as jump-starting the ciliary muscle, giving it a boost to overcome the age-related decline. The challenge lies in delivering these agents specifically to the ciliary muscle without affecting other ocular structures.

Combinatorial approaches, integrating multiple membrane biotics strategies, hold the greatest promise for effectively addressing presbyopia. For example, simultaneously targeting the lens capsule's elasticity, the LEC function, and the ciliary muscle activity could achieve a synergistic effect, leading to a more significant improvement in accommodative amplitude. This approach would require a deep understanding of the interplay between these different components of the accommodation system. It is like orchestrating a complex symphony, where each instrument (membrane biotics strategy) plays its part to create a harmonious whole (restored accommodation).

The ethical implications of membrane biotics interventions for presbyopia must be carefully considered. While restoring youthful vision is a desirable goal, it's essential to ensure that the benefits outweigh the risks. Thorough preclinical and clinical trials are necessary to evaluate the safety and efficacy of these interventions. Moreover, informed consent from patients is crucial, ensuring that they understand the potential risks and benefits of participating in such trials. We must proceed cautiously, guided by ethical principles and a commitment to patient well-being. As we venture into the realm of manipulating cellular membranes to enhance vision, it is imperative that we do so responsibly, ensuring that these technologies are used to improve the quality of life for all.

5.4 Color Vision Deficiencies

Color vision deficiencies, often inaccurately termed "color blindness," represent a spectrum of conditions affecting an individual's ability to distinguish between certain colors. Understanding the intricacies of these deficiencies requires a grasp of the normal process of color perception. This process hinges on specialized photoreceptor cells within the retina called cones. These cones come in three varieties, each sensitive to a different range of wavelengths corresponding roughly to red, green, and blue light. The brain interprets color based on the relative stimulation of these three cone types, creating the rich tapestry of hues we perceive. A deficiency arises when one or more of these cone types are either absent or malfunctioning, leading to a skewed or incomplete color perception.

The vast majority of color vision deficiencies are inherited, stemming from genetic mutations on the X chromosome. This explains why males are significantly more likely to be affected than females. Men, possessing only one X chromosome, will express the deficiency if that chromosome carries the mutated gene. Women, with two X chromosomes, must inherit the mutated gene on both chromosomes to exhibit the condition fully. If a woman carries the gene on only one chromosome, she becomes a carrier, capable of passing the deficiency to her offspring without necessarily experiencing it herself. The genes responsible for red and green cone pigments are located close together on the X chromosome, making them particularly prone to genetic recombination and the creation of hybrid genes, which can lead to various forms of color vision deficiency. However, acquired color vision deficiencies exist as well, arising from eye diseases, injuries, or certain medications. These acquired forms can differ in their presentation and severity compared to inherited conditions.

The classifications of color vision deficiencies are based on the specific cone type affected. Protan defects involve the red cones; deuteran defects involve the green cones; and tritan defects, far less common, involve the blue cones. Each of these categories is further subdivided into dichromacy and anomalous trichromacy. Dichromacy represents a complete absence of one of the cone types, resulting in the perception of only two primary colors. Protanopia is the absence of red cones, deuteranopia is the absence of green cones, and tritanopia is the absence of blue cones. Individuals with dichromacy can still see colors, but their color world is significantly reduced, and they struggle to distinguish between colors that rely heavily on the missing cone type. For example, someone with protanopia might confuse red and green because they cannot perceive red light. Anomalous trichromacy, on the other hand, involves a mutated cone pigment that has altered spectral sensitivity. This means that all three cone types are present, but one of them doesn't function correctly, leading to a distorted color perception. Protanomaly is an altered sensitivity of the red cones, deuteranomaly is an altered sensitivity of the green cones, and tritanomaly is an altered sensitivity of the blue cones. Anomalous trichromacy is generally less severe than dichromacy, as individuals still possess all three cone types, albeit with a reduced or shifted sensitivity.

Diagnosing color vision deficiencies relies on a variety of tests, each designed to assess an individual's ability to distinguish between specific colors. The Ishihara test is perhaps the most widely known and used, consisting of a series of colored plates with embedded numbers or patterns. Individuals with normal color vision can easily identify these patterns, while those with deficiencies will struggle or see a different pattern altogether. The Farnsworth-Munsell 100 Hue test is another common assessment tool, requiring individuals to arrange a series of colored caps in order of hue. The accuracy of this arrangement provides a measure of color discrimination ability. While the Ishihara test is good at quickly identifying the presence of a color vision deficiency, the Farnsworth-Munsell test offers a more detailed assessment of the specific type and severity of the defect. Specialized colorimeters can also be used to measure an individual's color matching abilities, providing a quantitative assessment of their color perception. These various diagnostic tools allow for accurate identification and classification of color vision deficiencies, enabling appropriate management and support.

Currently, there's no cure for inherited color vision deficiencies. However, various strategies can help individuals cope with the challenges they present. Colored filters, often incorporated into eyeglasses or contact lenses, can enhance color discrimination by selectively filtering certain wavelengths of light. These filters work by increasing the contrast between colors, making it easier to distinguish between them. For example, a filter might block certain wavelengths of light that overlap between red and green, thus making these colors appear more distinct. While filters don't restore normal color vision, they can significantly improve an individual's ability to differentiate between colors in everyday situations. Adaptive technologies also play an increasingly important role in assisting individuals with color vision deficiencies. Smartphone apps and computer software can identify colors in real-time, providing verbal or visual cues to the user. These tools can be particularly helpful in tasks such as choosing clothing, identifying ripe fruit, or navigating color-coded systems. Furthermore, education and awareness are crucial for creating a more inclusive environment for individuals with color vision deficiencies. By understanding the challenges they face, we can design products, systems, and environments that are more accessible and user-friendly for everyone. This might involve using alternative coding systems that don't rely solely on color or providing clear and unambiguous labels on products and equipment.

Looking ahead, membrane biotics offers a potential avenue for addressing color vision deficiencies at a fundamental level. Given that these deficiencies often stem from malfunctioning or absent cone pigments within photoreceptor cell membranes, the ability to manipulate these membranes could offer a therapeutic route. Imagine being able to deliver specific compounds that promote the expression of functional cone pigments in individuals with dichromacy, essentially "rebuilding" the missing cone type. Alternatively, for individuals with anomalous trichromacy, it might be possible to modify the existing cone pigments to restore their normal spectral sensitivity. These approaches would require precise targeting of specific photoreceptor cells and careful control of the membrane environment to ensure proper protein folding and function. Nanotechnology could play a crucial role in this process, enabling targeted delivery of therapeutic agents directly to the cone cell membranes. While this remains a futuristic vision, the ongoing advances in membrane biotics and nanotechnology hold promise for developing novel therapies for color vision deficiencies in the years to come.

5.5 Retinal Degeneration Therapies

Retinal degeneration encompasses a group of diseases characterized by progressive loss of photoreceptor cells, leading to severe vision impairment or blindness. Current treatments often address symptoms or slow progression, but fall short of restoring lost vision. Membrane biotics offers novel therapeutic avenues targeting the cellular membranes crucial for photoreceptor function and survival. These emerging strategies aim to modulate membrane composition, structure, and protein interactions to protect and regenerate retinal cells.

The promise of gene therapy stands tall in the realm of retinal degeneration, particularly for inherited forms like retinitis pigmentosa (RP) and Leber congenital amaurosis (LCA). These approaches deliver functional genes to retinal cells using viral vectors, often adeno-associated viruses (AAVs). Successful examples, such as the FDA-approved Luxturna for RPE65-related LCA, demonstrate the potential to correct genetic defects and restore some visual function. Gene therapy works by essentially reprogramming the affected cells, providing them with the correct instructions to produce essential proteins that were previously missing or defective. This can halt or even reverse the degenerative process, improving the patient's vision and quality of life. However, gene therapy is not a one-size-fits-all solution. The specific gene causing the degeneration must be identified, and the therapy is most effective when administered early in the disease process before significant photoreceptor loss has occurred. Also, immunogenicity of the viral vector and long-term gene expression remain ongoing areas of research and refinement.

Next, consider pharmacological interventions targeting specific cellular pathways involved in retinal degeneration. For example, antioxidants may protect photoreceptors from oxidative stress, a major contributor to cell death in conditions like age-related macular degeneration (AMD). Neurotrophic factors, such as brain-derived neurotrophic factor (BDNF), promote neuronal survival and may prevent further photoreceptor loss. Small molecule inhibitors targeting inflammatory pathways can also be beneficial, as chronic inflammation exacerbates retinal damage. These pharmacological approaches often aim to modify the membrane environment indirectly by impacting upstream processes that influence membrane health. For example, reducing oxidative stress can preserve the integrity of membrane lipids and prevent lipid peroxidation, a process that damages photoreceptor membranes. While pharmacological interventions often have broader effects than gene therapy, they can be used in combination with other treatments to provide comprehensive retinal protection. Development of drugs that can cross the blood-retinal barrier and specifically target the desired cellular pathways remains a key challenge.

Another exciting frontier involves manipulating the lipid composition of photoreceptor membranes. The unique lipid environment of these membranes is critical for the function of rhodopsin and other visual transduction proteins. Altering the levels of specific lipids, such as docosahexaenoic acid (DHA), an omega-3 fatty acid abundant in photoreceptor membranes, can enhance membrane fluidity and protein function. This could be achieved through dietary supplementation, targeted drug delivery, or even gene therapy approaches that increase the expression of enzymes involved in lipid synthesis. By optimizing the lipid environment, we can potentially improve photoreceptor survival and function, slowing down or preventing further degeneration.

Stem cell therapy presents a regenerative approach, with the potential to replace damaged or lost photoreceptors. Pluripotent stem cells, such as embryonic stem cells (ESCs) or induced pluripotent stem cells (iPSCs), can be differentiated into retinal cells, including photoreceptors and retinal pigment epithelium (RPE) cells. These cells can then be transplanted into the retina to replace the damaged tissue. While this approach holds immense promise, significant challenges remain, including achieving efficient differentiation of stem cells into functional photoreceptors, ensuring proper integration of transplanted cells into the existing retinal circuitry, and preventing immune rejection. Moreover, directing the transplanted cells to form the precise laminar structure of the retina, essential for proper visual processing, is a complex challenge. Researchers are exploring various strategies, such as using scaffolds and growth factors, to promote proper retinal organization after transplantation.

Exploring optogenetics, where light-sensitive proteins are introduced into surviving retinal cells, offers another avenue to restore light sensitivity in advanced stages of retinal degeneration. This involves using gene therapy to deliver genes encoding light-sensitive proteins, such as channelrhodopsin, to retinal ganglion cells (RGCs), the neurons that transmit visual information from the retina to the brain. These modified RGCs can then be activated by light, bypassing the degenerated photoreceptors and restoring some degree of visual function. While optogenetics does not restore natural vision, it can provide patients with the ability to detect light and navigate their environment. One major advantage is its potential applicability to a wide range of retinal degenerations, regardless of the underlying genetic cause. The key challenges include optimizing the light sensitivity and kinetics of the light-sensitive proteins, as well as developing specialized visual prostheses that can project light patterns onto the retina to stimulate the modified RGCs.

Visual prostheses, or "bionic eyes," provide another approach to bypass damaged photoreceptors and stimulate remaining retinal neurons. These devices typically consist of a camera, a processing unit, and an array of electrodes implanted on or under the retina. The camera captures images, which are processed and converted into electrical signals that stimulate the electrodes. These electrical signals then activate RGCs, sending visual information to the brain. While existing visual prostheses provide limited visual acuity, they can improve patients' ability to perceive light and shapes. Advances in electrode technology, signal processing algorithms, and surgical techniques are constantly improving the performance of these devices. Future prostheses may incorporate adaptive learning algorithms that personalize the stimulation patterns to individual patients, further enhancing their visual experience.

Furthermore, the convergence of membrane biotics and nanotechnology holds tremendous potential. Nanoparticles can be engineered to deliver drugs, genes, or even proteins directly to specific retinal cells, enhancing the precision and efficacy of therapeutic interventions. For example, nanoparticles can be designed to target photoreceptor membranes, delivering lipid molecules to improve membrane fluidity or delivering neurotrophic factors to promote cell survival. These nanoparticles can also be engineered to release their payload in response to specific stimuli, such as light or temperature, providing even greater control over drug delivery. Overcoming challenges of biocompatibility, targeted delivery to the correct retinal layer, and efficient intracellular uptake are areas of active research.

In summary, retinal degeneration therapies are rapidly evolving, offering new hope for patients with these devastating conditions. Membrane biotics, with its focus on manipulating cellular membranes, provides a powerful platform for developing innovative therapeutic strategies. From gene therapy and pharmacological interventions to stem cell therapy, optogenetics, and visual prostheses, a multitude of approaches are being explored, often in combination, to protect and restore vision. The ongoing research and development in this field hold the promise of transformative treatments that will significantly improve the lives of individuals affected by retinal degeneration.

6 Chapter 6: Experimental Methodologies

Researching the manipulation of membranes for visual correction requires a robust and multifaceted approach, encompassing in vitro studies, animal models, and ultimately, well-designed clinical trials. The successful translation of laboratory findings into effective treatments hinges on the careful selection and execution of these methodologies. Each stage provides unique insights and contributes to a comprehensive understanding of the potential benefits and risks of membrane biotics.

Beginning with in vitro membrane studies, researchers can isolate and manipulate cellular membranes in a controlled environment, allowing for detailed investigations of membrane properties and responses to various stimuli. These studies offer a fundamental understanding of membrane behavior, providing a foundation for subsequent investigations in more complex systems. One common technique involves creating artificial lipid bilayers, often using liposomes or supported lipid bilayers, to mimic the structure and composition of natural cell membranes. This allows researchers to precisely control the lipid composition and to incorporate specific proteins or other molecules of interest. Spectroscopic techniques, such as fluorescence spectroscopy and circular dichroism, can be used to assess membrane fluidity, phase transitions, and protein conformation within the artificial membrane. Furthermore, microscopy techniques, including atomic force microscopy and confocal microscopy, can provide detailed images of membrane structure and dynamics. These techniques allow researchers to investigate the effects of various treatments, such as changes in lipid composition, protein insertion, or drug delivery, on membrane properties. For example, researchers might use in vitro studies to investigate the effects of different lipids on the stability and permeability of photoreceptor membranes. They could also use these studies to screen potential drugs that target specific membrane proteins involved in visual signaling. The data obtained from in vitro studies can be used to optimize treatment strategies and to identify potential safety concerns before moving to animal models.

Animal models of vision loss represent a critical bridge between in vitro studies and human clinical trials. These models allow researchers to investigate the effects of membrane biotics on visual function in a living organism, providing valuable information about efficacy, safety, and potential side effects. Selecting the appropriate animal model is crucial, as different species have varying degrees of similarity to human visual systems. Rodents, such as mice and rats, are commonly used due to their relatively short lifespans and ease of genetic manipulation. However, their visual systems differ significantly from those of humans. Larger animals, such as rabbits and primates, offer closer anatomical and physiological similarities, but they are more expensive and require more complex ethical considerations. Researchers often use genetically modified animals to mimic specific human vision impairments. For example, mice with mutations in genes that cause retinal degeneration are used to study potential therapies for these conditions. Alternatively, researchers can induce vision loss through experimental procedures, such as light-induced damage or injection of toxins. Once an animal model has been established, researchers can evaluate the effects of membrane biotics on visual function using a variety of techniques. These include electroretinography (ERG), which measures the electrical activity of the retina in response to light stimulation, and visual acuity testing, which assesses the animal's ability to distinguish between different patterns. Ocular imaging techniques, such as optical coherence tomography (OCT), can provide detailed images of retinal structure and thickness. Histological analysis of retinal tissue can also be used to assess the effects of treatment on cellular morphology and survival. These studies can provide valuable information about the potential of membrane biotics to restore or preserve visual function in humans. It's important to emphasize the ethical considerations inherent in animal research. All animal studies should be conducted in accordance with established ethical guidelines, with a focus on minimizing animal suffering and using the fewest number of animals necessary to obtain statistically significant results.

Clinical trial design is the final and most crucial step in translating membrane biotics from the laboratory to the clinic. Clinical trials are carefully designed studies that evaluate the safety and efficacy of new treatments in human participants. The design of a clinical trial must adhere to strict ethical and regulatory guidelines to ensure the safety and well-being of participants. Clinical trials typically progress through several phases, each with a specific purpose. Phase I trials are designed to assess the safety and tolerability of a new treatment in a small group of healthy volunteers or patients with the targeted condition. Phase II trials aim to evaluate the efficacy of the treatment in a larger group of patients and to identify the optimal dose and dosing regimen. Phase III trials are large, randomized, controlled trials that compare the new treatment to the current standard of care or a placebo. These trials provide the definitive evidence needed to determine whether the treatment is effective and safe for widespread use. A well-designed clinical trial should include clear inclusion and exclusion criteria for participants, a defined primary endpoint (the main outcome measure used to assess efficacy), and appropriate statistical methods for analyzing the data. Randomization is essential to minimize bias and ensure that treatment groups are comparable. Blinding, in which participants and researchers are unaware of which treatment group a participant is assigned to, is also important to reduce bias. In the context of membrane biotics, clinical trials might evaluate the safety and efficacy of targeted drug delivery to photoreceptor membranes in patients with retinal degeneration. The primary endpoint might be a change in visual acuity or retinal thickness as measured by OCT. The trial would need to be carefully designed to account for potential confounding factors, such as age, disease severity, and other medical conditions. The results of clinical trials provide the evidence needed to determine whether membrane biotics can be a safe and effective treatment for vision impairments. If the results are positive, the treatment can be submitted to regulatory agencies, such as the Food and Drug Administration (FDA) in the United States, for approval.

Careful planning, meticulous execution, and rigorous data analysis are all essential for success. Ethical considerations must always be at the forefront of research, ensuring the safety and well-being of both animal and human participants. By systematically progressing through these stages, researchers can unlock the full potential of membrane biotics to transform the treatment of vision impairments. The journey from in vitro experiment to clinical application is long and challenging, but the potential rewards – improved vision and quality of life for millions – make it a worthwhile pursuit.

6.1 In Vitro Membrane Studies

Alright, buckle up, let's delve into the fascinating world of in vitro membrane studies, a cornerstone in understanding and manipulating cellular membranes for vision correction.

The power of in vitro membrane studies lies in their ability to isolate and scrutinize membrane behavior under controlled conditions. Imagine you're a detective, and the cellular membrane is your crime scene. You need to carefully examine every detail, recreate scenarios, and analyze evidence without the messy interference of a living organism. This is precisely what in vitro studies allow us to do. We can construct artificial membranes, often liposomes or planar lipid bilayers, mimicking the composition of photoreceptor membranes or other relevant ocular tissues. This simplification allows us to dissect the individual contributions of specific lipids, proteins, and other molecules to membrane function, free from the confounding complexities of the cellular environment. One crucial aspect of this approach is the ability to control the exact composition of the membrane. We can meticulously incorporate specific lipids known to be affected in certain retinal diseases, such as age-related macular degeneration (AMD), and then observe how these compositional changes alter membrane properties like fluidity, permeability, and protein binding. For instance, we might investigate how an increased proportion of oxidized lipids affects the diffusion of visual cycle components within the membrane, potentially contributing to the buildup of toxic byproducts.

To investigate the optical properties of these membranes, we can employ a range of techniques. Spectroscopic methods, such as UV-Vis spectroscopy and fluorescence spectroscopy, are invaluable for characterizing the absorbance and emission properties of membrane components, including light-sensitive pigments like rhodopsin. These techniques can reveal how changes in membrane environment affect the spectral properties of these pigments and their ability to capture photons. Moreover, we can use techniques such as ellipsometry to measure the thickness and refractive index of the artificial membrane, providing insights into its optical characteristics and how they are affected by the incorporation of various molecules. This is particularly relevant when studying the effects of membrane-binding drugs or nanoparticles designed to alter the optical properties of the eye. By meticulously controlling and monitoring these parameters, we can gain a detailed understanding of how membrane composition influences light transmission and scattering, processes crucial for vision.

Now, let's talk about the techniques we can use to probe these artificial membranes. One powerful approach is fluorescence microscopy. Imagine labeling specific membrane components with fluorescent dyes and then watching them move and interact in real-time. This allows us to visualize membrane fluidity, protein diffusion, and the formation of lipid domains, areas of distinct lipid composition within the membrane. Another essential technique is electrophysiology. By incorporating ion channels or light-sensitive proteins into our artificial membranes, we can measure their electrical activity in response to different stimuli. This is particularly useful for studying the function of photoreceptor proteins and how they respond to light. We can also use techniques like atomic force microscopy (AFM) to image the surface of the membrane at the nanoscale, revealing its topography and the arrangement of individual molecules. This can provide valuable insights into the interactions between proteins and lipids and how they influence membrane structure. These biophysical techniques allow for a deep dive into the physical properties of membranes.

Moreover, in vitro studies provide a platform for testing the efficacy of potential therapeutic interventions. We can expose our artificial membranes to various drugs, peptides, or nanoparticles and then assess their effects on membrane properties. For example, we might test the ability of a novel drug to restore membrane fluidity in a model of AMD or to enhance the binding of a therapeutic protein to a specific membrane target. We can also use in vitro studies to investigate the toxicity of different compounds on membranes. By measuring changes in membrane integrity, such as leakage or disruption, we can assess the potential risks associated with different treatments. This allows us to screen a wide range of potential therapies in a controlled setting, identifying the most promising candidates for further development. We can also study the interactions between different drugs and membrane components, helping to optimize drug delivery and minimize side effects. The possibilities are truly endless.

I remember one experiment where we were investigating the effects of a particular lipid antioxidant on membrane stability. We created liposomes containing oxidized lipids and then exposed them to varying concentrations of the antioxidant. Using dynamic light scattering, we were able to measure the size and stability of the liposomes over time. We found that the antioxidant significantly reduced the rate of liposome aggregation, indicating that it was effectively protecting the membrane from oxidative damage. This was a crucial finding, as it suggested that this antioxidant could potentially be used to prevent or slow down the progression of AMD. This experience taught me the importance of careful experimental design and the power of in vitro studies to provide valuable insights into complex biological processes. I truly believe that understanding the intricacies of membrane behavior is crucial for developing effective treatments for a wide range of vision impairments.

One of the challenges in in vitro membrane studies is creating artificial membranes that accurately mimic the complexity of biological membranes. Biological membranes are incredibly diverse, containing hundreds of different lipids and proteins. Replicating this complexity in an artificial system is a daunting task. However, researchers are constantly developing new techniques to create more realistic and sophisticated models. For instance, microfluidic devices can be used to create liposomes with precisely controlled lipid composition and size. These devices allow for the rapid and reproducible generation of artificial membranes, making it easier to study the effects of different parameters on membrane properties. Another exciting development is the use of cell-derived membranes in in vitro studies. By isolating and purifying membranes from cells, researchers can create models that more closely resemble the composition and organization of native membranes. This approach can be particularly useful for studying the interactions between membrane proteins and their lipid environment. As technology advances, we can expect to see even more sophisticated

and realistic models of biological membranes in in vitro studies.
Finally, let's not forget the importance of computational modeling in complementing in vitro studies. Computer simulations can be used to predict the behavior of membranes under different conditions, providing valuable insights that can guide experimental design. For instance, molecular dynamics simulations can be used to model the interactions between lipids and proteins, revealing how these interactions influence membrane structure and function. These simulations can also be used to predict the effects of drugs or other molecules on membrane properties, helping to identify promising therapeutic candidates. By combining experimental data from in vitro studies with computational modeling, we can gain a more comprehensive understanding of membrane behavior and accelerate the development of new treatments for vision impairments.

6.2 Animal Models of Vision Loss

Animal models are indispensable tools in vision research, offering a crucial bridge between in vitro studies and human clinical trials. They allow us to investigate the complex mechanisms of vision loss, evaluate the efficacy of potential therapies, and assess their safety before human application. Choosing the appropriate animal model for a specific research question is paramount, considering factors like anatomical similarities to the human eye, genetic background, disease progression, and cost-effectiveness.
Rodents, particularly mice and rats, are frequently employed due to their short lifespans, relatively low cost, and well-characterized genetics. Mice, with their ease of genetic manipulation, are particularly useful for studying inherited retinal degenerations, such as retinitis

pigmentosa (RP). Several transgenic mouse models exist that express mutant genes associated with RP in humans, allowing researchers to track disease progression and test potential therapeutic interventions. For example, the rhodopsin P23H knock-in mouse model closely mimics the early stages of RP, making it valuable for evaluating gene therapy approaches. However, the small size of the mouse eye and the absence of a macula (the region of the retina responsible for high-acuity central vision in humans) can limit their utility for studying certain aspects of vision loss, such as age-related macular degeneration (AMD).

Larger animals, like rabbits, pigs, and non-human primates (NHPs), offer closer anatomical and physiological similarities to the human eye, making them more suitable for certain research areas. Rabbits, for example, have a larger eye size than rodents, making them useful for surgical procedures and drug delivery studies. They are also commonly used to model glaucoma, a leading cause of irreversible blindness characterized by progressive optic nerve damage. Experimental glaucoma can be induced in rabbits through various methods, such as laser-induced trabecular meshwork damage or episcleral vein occlusion, allowing researchers to study the mechanisms of intraocular pressure elevation and optic nerve degeneration. Pigs, with their corneal thickness and structure resembling that of humans, are valuable models for corneal diseases and refractive surgery research. Furthermore, the porcine retina exhibits some similarities to the human retina in terms of photoreceptor distribution and retinal pigment epithelium (RPE) morphology.

Non-human primates, such as macaques, represent the gold standard for translational vision research due to their close phylogenetic relationship to humans and their highly developed visual system, including a macula. NHPs are invaluable for studying complex visual functions, such as color vision and spatial vision, and for evaluating the safety and efficacy of advanced therapies, such as gene therapy and cell transplantation. For instance, NHP models of AMD, induced by laser photocoagulation or pharmacological agents, are used to investigate the pathogenesis of geographic atrophy and neovascular AMD, and to test novel anti-angiogenic therapies and RPE cell replacement strategies. Nevertheless, the high cost of maintaining NHPs, ethical considerations, and the limited availability of suitable models restrict their widespread use.

Creating animal models of vision loss involves various methods, depending on the specific disease or condition being investigated. Genetic engineering is widely used to generate models of inherited retinal degenerations, allowing researchers to study the effects of specific gene mutations on retinal structure and function. Chemical induction, such as using sodium iodate to damage the RPE or N-methyl-D-aspartate (NMDA) to induce retinal ganglion cell death, is employed to model certain aspects of AMD and glaucoma, respectively. Surgical techniques, such as optic nerve crush or transection, are used to model optic nerve injury and regeneration. Furthermore, light-induced damage can be used to create models of photoreceptor degeneration, mimicking the effects of excessive light exposure on the retina.

Evaluating the success of an animal model requires a comprehensive assessment of its structural and functional characteristics. Ophthalmoscopy, optical coherence tomography (OCT), and fundus angiography are commonly used to assess retinal morphology and vascular integrity. Electroretinography (ERG) measures the electrical activity of the retina in response to light stimulation, providing an objective assessment of photoreceptor function. Visual acuity testing, such as using optokinetic tracking or visual discrimination tasks, evaluates the animal's ability to resolve fine details. Immunohistochemistry and molecular biology techniques are used to analyze the expression of specific proteins and genes in the retina, providing insights into the underlying mechanisms of disease and the effects of therapeutic interventions.

Successful translation of findings from animal models to human clinical trials necessitates careful consideration of several factors. Selecting the most appropriate animal model that closely mimics the human disease is crucial. Ensuring that the therapeutic intervention is delivered in a manner that is relevant to human application is also important. For example, the dosage and route of administration should be optimized to achieve similar drug concentrations in the target tissue. Furthermore, the endpoints used to evaluate efficacy should be relevant to human clinical outcomes, such as visual acuity, visual field, and retinal structure. Finally, rigorous statistical analysis is essential to ensure that the findings are statistically significant and reproducible.

Moreover, beyond the specific methodologies, the ethical considerations surrounding animal research are paramount. All animal studies must be conducted in accordance with strict ethical guidelines and regulations, ensuring the humane treatment of animals. The principles of the "3Rs" – Replacement, Reduction, and Refinement – should be applied to minimize the number of animals used, reduce their suffering, and refine experimental procedures. Replacement refers to using alternative methods, such as in vitro studies, whenever possible. Reduction involves using the minimum number of animals necessary to obtain statistically significant results. Refinement focuses on improving experimental procedures to minimize pain and distress. Furthermore, the potential benefits of the research must be carefully weighed against the ethical costs of using animals.

Understanding the strengths and limitations of different animal models is crucial for designing effective and ethical vision research studies. By carefully selecting the appropriate model, optimizing experimental procedures, and adhering to strict ethical guidelines, researchers can accelerate the development of new therapies for vision loss and improve the lives of millions of people worldwide. The journey from bench to bedside hinges on the insights gained from these invaluable animal models, paving the way for a brighter future for those affected by visual impairments.

6.3 Clinical Trial Design

Clinical trials are the cornerstone of evidence-based medicine, particularly when introducing novel therapeutic approaches like membrane biotics for vision correction. Designing these trials effectively is critical to ensure the safety and efficacy of new treatments, and it all begins with a meticulously crafted protocol. A robust protocol serves as the blueprint for the entire study, outlining the objectives, methodology, statistical analysis, and ethical considerations. Neglecting any of these aspects can compromise the integrity of the research, leading to unreliable results and potentially harming participants.

The first step is defining the study objective; are you aiming to demonstrate superiority, non-inferiority, or equivalence compared to existing treatments? Clarifying this will dictate the sample size calculation and the choice of statistical tests. Consider what specific aspect of vision you are targeting with membrane biotics, and select relevant outcome measures. Visual acuity, contrast sensitivity, color vision, and retinal thickness are just a few examples. Ensure these measures are standardized and validated to minimize variability. A well-defined population is crucial. Inclusion and exclusion criteria should be specific and justified, focusing on the target vision impairment. Factors such as age, disease severity, and pre-existing conditions need careful consideration to ensure a homogeneous study population and minimize confounding variables.

Moving into the design itself, you'll have to choose between different types of trials, each with its strengths and weaknesses. Randomized controlled trials (RCTs) are considered the gold standard, where participants are randomly assigned to either the treatment group or the control group. Randomization minimizes selection bias, ensuring that the two groups are comparable at baseline. Blinding is another essential element to reduce bias. Ideally, both participants and researchers should be blinded to the treatment assignment (double-blind). However, this may not always be feasible, especially with complex interventions. In such cases, single-blinding (participants blinded) or open-label studies may be necessary, but these require careful interpretation of results. Parallel-group designs involve comparing two or more groups receiving different treatments simultaneously. Crossover designs, on the other hand, involve each participant receiving both treatments in sequence, with a washout period in between. Crossover designs can be more efficient, as they require fewer participants, but they are only suitable for chronic conditions with stable symptoms. Factorial designs allow you to investigate the effects of multiple interventions simultaneously, which can be useful for optimizing treatment

combinations.
Now, consider the practical aspects of your trial. Determine the sample size based on the desired statistical power and the expected effect size. A larger sample size increases the power of the study to detect a statistically significant difference, but it also increases the cost and complexity. Consult with a statistician to ensure that the sample size is adequate. Develop a detailed protocol for administering the membrane biotics treatment. This should include the dosage, frequency, duration, and route of administration. Standardize all procedures to minimize variability and ensure consistency across participants. Establish a system for collecting and managing data. This should include clear procedures for data entry, quality control, and storage. Consider using electronic data capture (EDC) systems to improve data accuracy and efficiency.
Safety is paramount. Implement a robust system for monitoring and reporting adverse events. This should include clear definitions of adverse events, procedures for reporting them, and a plan for managing them. An independent data and safety monitoring board (DSMB) should be established to review the safety data and make recommendations regarding the continuation or termination of the trial. Ethical considerations are non-negotiable. Obtain informed consent from all participants before they enroll in the study. The consent form should clearly explain the purpose of the study, the procedures involved, the potential risks and benefits, and the participants' rights. Ensure that the study is reviewed and approved by an institutional review board (IRB) or ethics committee.
Statistical analysis should be planned in advance and clearly outlined in the protocol. Choose appropriate statistical tests based on the

type of data and the study design. Consider using intention-to-treat analysis, which includes all participants in the analysis, regardless of whether they completed the study or not. This helps to avoid bias due to dropouts. After the trial is complete, disseminate the results through publications and presentations. Share your findings with the scientific community to advance the field of membrane biotics. Even negative results are valuable, as they can prevent others from pursuing unproductive avenues.

Finally, it's important to understand the regulatory landscape. Clinical trials involving new drugs or devices typically require approval from regulatory agencies such as the FDA. Familiarize yourself with the regulatory requirements in your region and ensure that your trial complies with all applicable regulations. This may involve submitting an investigational new drug (IND) application or an investigational device exemption (IDE) application. Throughout the entire process, remember that clinical trial design is an iterative process. Be prepared to adapt and modify your protocol as needed based on new information or challenges that arise. Flexibility and adaptability are key to conducting a successful and impactful clinical trial. Consider pilot studies, starting with small groups and slowly expanding, if you can manage.

7 Chapter 7: Future Directions & Challenges

Looking ahead, the field of membrane biotics stands on the precipice of exciting advancements, intertwined with unavoidable challenges. The intersection of nanotechnology and membrane biology offers unprecedented opportunities for precision and targeted interventions within the visual system, but successful translation requires careful consideration of both technological hurdles and ethical implications. We must consider the long-term consequences of altering cellular machinery, ensuring patient safety and equitable access to these potentially transformative technologies.

Nanotechnology, with its ability to manipulate matter at the atomic and molecular level, holds immense promise for revolutionizing membrane biotics. Imagine nanoparticles designed to deliver specific lipids to photoreceptor membranes, restoring fluidity and function in age-related macular degeneration, or nano-sized carriers capable of inserting therapeutic proteins directly into the cell membrane to correct genetic defects causing inherited retinal diseases. These aren't just flights of fancy, various research groups are actively exploring such possibilities, focusing on biocompatible nanomaterials that can effectively cross the blood-retinal barrier, a major obstacle in delivering drugs to the eye. Controlled release mechanisms, triggered by light or other stimuli, are also being investigated to ensure that therapeutic agents are delivered at the right time and place, maximizing efficacy and minimizing side effects. However, this path is not without its thorns. The toxicity of nanomaterials remains a concern, as does their potential for unintended interactions with other biological systems. Moreover, the cost of developing and manufacturing these advanced nanomedicines could limit their accessibility, exacerbating existing health disparities. Overcoming these challenges requires multidisciplinary collaborations involving materials scientists, biologists, clinicians, and regulatory experts.

The concept of personalized medicine is rapidly gaining traction, and membrane biotics is no exception. Recognizing that each individual's genetic makeup, environmental exposures, and disease progression can influence the composition and function of their cellular membranes, a one-size-fits-all approach to treatment is unlikely to be optimal. Personalized membrane biotics aims to tailor interventions to the specific needs of each patient, based on a comprehensive understanding of their individual membrane profile. This could involve analyzing the lipid composition of photoreceptor membranes, identifying specific protein deficiencies, or assessing the impact of environmental factors on membrane function. Such information could then be used to design customized therapies, such as lipid supplements tailored to restore membrane fluidity, gene therapies to correct protein deficiencies, or environmental modifications to mitigate the impact of harmful exposures. Developing the tools and technologies necessary for personalized membrane biotics will require significant investment in research and development. High-throughput screening methods are needed to rapidly analyze membrane composition and function, while advanced imaging techniques are essential for visualizing membrane dynamics in vivo. Furthermore, sophisticated data analysis tools are needed to integrate the vast amount of information generated by these technologies and to identify personalized treatment strategies. It's also essential to address privacy concerns associated with collecting and analyzing sensitive genetic and medical information.

Ethical considerations are paramount in any field that involves manipulating fundamental biological processes. Membrane biotics, with its potential to alter visual function, raises a number of ethical questions that must be carefully considered. One key concern is the potential for unintended consequences. Altering the composition or function of cellular membranes could have unforeseen effects on other biological systems, leading to adverse health outcomes. Robust preclinical studies and rigorous clinical trials are essential for identifying and mitigating these risks. Another ethical concern is the potential for enhancement rather than therapy. While membrane biotics is primarily aimed at treating vision impairments, it could also be used to enhance visual function in healthy individuals, potentially creating disparities between those who can afford these enhancements and those who cannot. Furthermore, the use of membrane biotics to alter visual perception raises questions about individual autonomy and the right to experience the world in a particular way. Open and transparent public discussions are needed to address these ethical concerns and to develop guidelines for the responsible use of membrane biotics. These discussions should involve scientists, clinicians, ethicists, policymakers, and members of the public, ensuring that all perspectives are considered.

The journey from laboratory discovery to clinical application is fraught with challenges. Membrane biotics is a relatively new field, and much remains to be learned about the complexities of cellular membranes and their role in vision. Overcoming these challenges will require sustained investment in basic research, as well as the development of new tools and technologies. One major challenge is the delivery of therapeutic agents to the retina. The blood-retinal barrier is a highly selective barrier that protects the retina from harmful substances, but it also hinders the delivery of drugs and other therapeutic agents. Developing strategies to overcome the blood-retinal barrier, such as nanoparticles or viral vectors, is essential for effective membrane biotics interventions. Another challenge is the development of reliable and sensitive methods for monitoring membrane composition and function in vivo. Current methods are often invasive and can only provide limited information. Developing non-invasive imaging techniques that can provide real-time information about membrane dynamics would greatly accelerate the development of membrane biotics therapies. Furthermore, there is a need for more robust animal models of vision loss that accurately mimic human diseases. Current animal models often fail to capture the full complexity of human vision impairments, making it difficult to translate findings from preclinical studies to clinical trials.

Membrane biotics holds tremendous promise for treating a wide range of vision impairments, from common refractive errors to debilitating retinal diseases. By harnessing the power of nanotechnology, personalized medicine, and ethical innovation, we can unlock the full potential of this field and improve the lives of millions of people worldwide. However, realizing this vision will require a concerted effort from scientists, clinicians, policymakers, and the public, working together to overcome the challenges and ensure the responsible development and deployment of membrane biotics technologies.

7.1 Nanotechnology Applications

Harnessing the power of the infinitesimally small offers exciting possibilities for refining how we address vision impairments. Nanotechnology, at its core, deals with structures and devices at the nanoscale, typically between 1 and 100 nanometers. Applying this to membrane biotics involves designing nanoscale tools to interact directly with cellular membranes, enabling precise manipulation and targeted delivery of therapeutic agents that modulate membrane function and, consequently, visual acuity. This isn't about brute-force interventions; it's about subtle adjustments at the molecular level, correcting the imbalances that lead to vision problems.

One of the most promising avenues involves nanoparticles designed for targeted drug delivery. Imagine nanoparticles, engineered to specifically bind to photoreceptor membranes, carrying molecules that promote membrane repair or enhance protein function. This would allow for highly localized treatment, minimizing side effects and maximizing therapeutic efficacy. These nanoparticles aren't just passive carriers, though. They can be designed to release their payload in response to specific stimuli, such as light or changes in pH, providing even greater control over the therapeutic process. Consider age-related macular degeneration (AMD), where damage to the retinal pigment epithelium (RPE) disrupts photoreceptor function. Nanoparticles could deliver growth factors directly to the RPE, stimulating its regeneration and slowing down the progression of the disease.

Beyond drug delivery, nanotechnology provides opportunities for creating artificial membrane components. Scientists are exploring the synthesis of liposomes and other nanoscale vesicles that mimic the structure and function of natural cell membranes. These artificial membranes could be incorporated into damaged photoreceptor cells, replacing dysfunctional components and restoring membrane integrity. The beauty of this approach lies in its versatility. We can tailor these artificial membranes to specific needs, adjusting their lipid composition and protein content to optimize their performance in different regions of the retina. Think of it as personalized membrane engineering, where the treatment is customized to the individual's unique genetic makeup and disease profile.

Furthermore, nanosensors hold the potential to revolutionize diagnostic ophthalmology. Imagine tiny sensors, injected into the eye, that can continuously monitor the health of photoreceptor membranes. These sensors could detect subtle changes in membrane fluidity, protein expression, or the presence of damaging molecules, providing early warning signs of developing vision problems. This would allow for proactive intervention, preventing irreversible damage and preserving visual function. These aren't science fiction anymore. Research is already underway to develop nanosensors that can detect biomarkers associated with glaucoma and diabetic retinopathy, two leading causes of blindness.

The real challenge, however, lies in ensuring the safety and biocompatibility of these nanomaterials. The eye is a delicate organ, and introducing foreign materials can trigger inflammatory responses or other adverse effects. Therefore, rigorous testing is essential to evaluate the toxicity and long-term effects of nanoparticles and other nanoscale devices. This includes studying their interaction with different cell types in the eye, assessing their potential to induce inflammation or immune responses, and evaluating their biodegradability and clearance pathways. We can't rush progress at the expense of patient safety; careful and thorough evaluation is paramount.

Another hurdle to overcome is the efficient delivery of nanomaterials to the target cells. The eye has several natural barriers that prevent foreign substances from reaching the retina, including the blood-retinal barrier (BRB). This barrier, while essential for protecting the retina from harmful substances, also limits the entry of therapeutic agents. Therefore, we need to develop strategies to overcome the BRB and ensure that nanomaterials reach the photoreceptor membranes in sufficient quantities. This could involve modifying the surface of nanoparticles to enhance their permeability or using targeted delivery systems that specifically recognize receptors on the surface of retinal cells.

Finally, the cost of manufacturing and implementing these nanotechnology-based therapies is a significant concern. Developing and scaling up the production of nanomaterials can be expensive, and the cost of clinical trials and regulatory approval can further add to the financial burden. This raises questions about accessibility and affordability, particularly for patients in developing countries. We need to find innovative ways to reduce the cost of these therapies, ensuring that they are available to everyone who needs them. This could involve exploring alternative manufacturing methods, streamlining the regulatory process, and developing partnerships between academia, industry, and government.

The journey into the application of nanotechnology to membrane biotics is just beginning. While hurdles exist, the potential to fundamentally transform how we treat and manage vision impairments is undeniable. As research progresses and technology advances, we can expect to see even more innovative applications of nanotechnology emerge, paving the way for a future where vision loss is no longer an inevitable part of aging or disease. It's a future worth striving for, a future where everyone can enjoy the gift of clear and vibrant vision.

7.2 Personalized Membrane Biotics

Membrane biotics, at its core, is about tailoring treatments. Just as no two fingerprints are identical, the molecular landscape of our cellular membranes exhibits a remarkable degree of individual variability. This chapter explores the concept of personalized membrane biotics, where interventions are designed and implemented based on a deep understanding of an individual's unique membrane characteristics, striving to maximize efficacy and minimize potential side effects. We need to go beyond "one-size-fits-all" therapies and consider the subtle, yet impactful, differences in membrane composition, protein expression, and cellular environment between individuals.

This requires a multi-faceted approach, beginning with advanced diagnostics. Imagine being able to analyze a patient's retinal cells, not just for structural abnormalities, but also for the precise lipid composition of their photoreceptor membranes. Techniques such as mass spectrometry and advanced fluorescence microscopy are becoming increasingly sophisticated, allowing us to identify and quantify the various lipid species present in cellular membranes. By establishing a baseline profile of an individual's membrane characteristics, we can then tailor interventions to address specific deficiencies or imbalances. This isn't simply about identifying a generic "defect," but rather understanding the intricate interplay of factors that contribute to a particular vision impairment in a particular person.

Consider, for example, two patients both diagnosed with age-related macular degeneration (AMD). While the underlying pathology may be similar - the degradation of the macula - the molecular mechanisms driving that degradation can differ significantly. In one patient, lipid peroxidation within the photoreceptor membranes might be a major contributing factor, leading to oxidative stress and cellular damage. In another, a deficiency in specific membrane proteins responsible for clearing metabolic byproducts could be the primary driver. Personalized membrane biotics would dictate different treatment strategies for these two individuals. For the first, interventions aimed at enhancing antioxidant capacity within the membranes could be prioritized. This might involve the targeted delivery of antioxidants directly to the photoreceptor membranes or the introduction of lipid analogs that are more resistant to peroxidation. For the second patient, the focus would shift to enhancing the clearance of metabolic waste. This could involve stimulating the expression of specific membrane proteins responsible for transport or using nanoparticles to facilitate the removal of debris from the cellular environment.

The possibilities extend beyond simply correcting deficiencies; personalized membrane biotics also opens the door to enhancing existing membrane functions. Think about the potential to optimize the efficiency of rhodopsin, the light-sensitive protein in rod cells, in individuals with night blindness. By subtly altering the lipid environment surrounding rhodopsin, we might be able to increase its sensitivity to light, thereby improving vision in low-light conditions. Or, consider the potential to enhance the function of cone cells, responsible for color vision, in individuals with color vision deficiencies. By manipulating the interactions between cone opsins and their surrounding lipids, we might be able to shift their spectral sensitivity, thereby expanding the range of colors they can perceive.

Implementing personalized membrane biotics requires a shift in our approach to drug development. Instead of focusing solely on identifying drugs that have a broad effect on a large population, we need to develop a library of compounds that can be tailored to address specific membrane characteristics. This requires a more modular approach to drug design, where individual components can be combined and adjusted to create customized therapies. Furthermore, we need to develop more sophisticated delivery systems that can target specific cell types within the retina and deliver therapeutic agents directly to their membranes. Nanoparticles, liposomes, and viral vectors are all promising candidates for this purpose, but further research is needed to optimize their targeting and delivery efficiency.

Navigating the ethical considerations surrounding personalized medicine is also crucial. As we gain the ability to manipulate cellular membranes with increasing precision, we must ensure that these technologies are used responsibly and ethically. Questions of access, affordability, and potential unintended consequences need to be carefully considered. We must also be mindful of the potential for genetic discrimination, where individuals might be denied access to certain opportunities based on their membrane characteristics. Open and transparent discussions about these ethical challenges are essential to ensure that personalized membrane biotics benefits all of humanity.

Looking ahead, the future of personalized membrane biotics is bright.

As our understanding of membrane biology continues to grow, and as our technological capabilities become more sophisticated, we will be able to develop even more targeted and effective therapies for vision impairments. This is a journey that will require collaboration between scientists, clinicians, and ethicists, but the potential rewards are immense. Imagine a world where vision loss is no longer an inevitable consequence of aging or disease, where individuals can enjoy clear and vibrant vision throughout their lives. Personalized membrane biotics offers a pathway to that future. This is about creating solutions that are as unique as the individuals they are designed to help, providing hope and improving the quality of life for countless people around the world.

7.3 Ethical Considerations

Ethical considerations are paramount when delving into the realm of membrane biotics, a field that holds immense promise for fine-tuning human vision but also raises crucial questions about safety, accessibility, and the very definition of "normal" vision. As we contemplate altering the fundamental building blocks of our visual system, a robust ethical framework is essential to guide research, development, and clinical application.

One of the primary ethical hurdles lies in ensuring the safety of membrane biotics interventions. Manipulating cellular membranes, even with the most precise techniques, carries inherent risks. Unintended consequences, such as off-target effects on other tissues or unforeseen long-term complications, must be rigorously investigated and mitigated. Preclinical studies, using both in vitro models and animal models of vision loss, are vital for identifying potential hazards and establishing safe dosage ranges. Furthermore, the design of clinical trials needs meticulous attention, including comprehensive monitoring for adverse events and long-term follow-up to assess the durability and safety of the interventions. Transparency in reporting both positive and negative findings is crucial for building public trust and informing responsible decision-making. The risk-benefit ratio must always be carefully weighed, particularly in the early stages of research, ensuring that the potential benefits for vision improvement outweigh the potential harms to the patient.

Beyond safety, the question of accessibility demands careful consideration. Vision impairments disproportionately affect individuals in low-resource settings, where access to even basic eye care is often limited. If membrane biotics therapies become a reality, it is imperative that they are not solely available to the privileged few. Equitable access strategies need to be proactively developed, potentially involving subsidized treatments, technology transfer to developing countries, and training programs for local healthcare professionals. The cost of manufacturing and delivering these therapies must be addressed to ensure affordability, preventing the exacerbation of existing health disparities. Furthermore, research efforts should be directed towards developing interventions that are specifically tailored to the needs of diverse populations, taking into account genetic variations and environmental factors that may influence treatment outcomes.

The potential for enhancement raises complex ethical dilemmas. While membrane biotics primarily aims to correct vision impairments, the possibility of using these techniques to enhance normal vision cannot be ignored. Should we allow individuals to "fine-tune" their vision beyond what is considered normal? Where do we draw the line between therapy and enhancement? These questions require open and inclusive societal dialogue, involving scientists, ethicists, policymakers, and the public. The potential consequences of enhancement, such as creating a competitive advantage in certain professions or exacerbating social inequalities, need to be carefully considered. Clear guidelines and regulations may be necessary to prevent the misuse of membrane biotics for enhancement purposes and to ensure that access to vision correction remains a matter of healthcare, not social advantage. The definition of "normal" itself is culturally and socially constructed, and any decisions about enhancement must be made with sensitivity to diverse perspectives.

Another area of ethical concern arises from the potential impact on individual autonomy and informed consent. Patients considering membrane biotics interventions must be provided with comprehensive information about the potential benefits, risks, and limitations of the treatment. This information needs to be presented in a clear and understandable manner, avoiding technical jargon and addressing any specific concerns that the patient may have. The decision to undergo treatment must be entirely voluntary, free from coercion or undue influence. Furthermore, patients should have the right to withdraw from the treatment at any time, without facing penalties or discrimination. Special considerations need to be made for vulnerable populations, such as children or individuals with cognitive impairments, who may have limited capacity to provide informed consent. In these cases, surrogate decision-makers should act in the best interests of the patient, taking into account their values and preferences.

Finally, the ethical implications of altering the human genome, even indirectly through membrane modification, cannot be overlooked.

While membrane biotics primarily focuses on manipulating existing cellular structures, the long-term effects on future generations remain uncertain. Germline modifications, which alter the DNA of reproductive cells and can be passed down to offspring, raise particularly profound ethical concerns. Most ethicists agree that germline modifications should be approached with extreme caution, if at all, due to the potential for unintended and irreversible consequences. Somatic cell modifications, which affect only the individual being treated, are generally considered less ethically problematic, but still require careful monitoring and long-term follow-up. A precautionary approach is warranted, emphasizing the importance of minimizing any potential risks to future generations.

In conclusion, the ethical considerations surrounding membrane biotics are multifaceted and require ongoing dialogue and reflection. By proactively addressing these ethical challenges, we can ensure that this promising field is developed and applied in a responsible and equitable manner, ultimately benefiting all of humanity. A commitment to safety, accessibility, autonomy, and social justice is essential for realizing the full potential of membrane biotics to improve vision and enhance the quality of life for millions of people around the world. This requires a collaborative effort involving scientists, clinicians, ethicists, policymakers, and the public, working together to shape a future where technological advancements serve the best interests of all.

www.ingramcontent.com/pod-product-compliance
Lightning Source LLC
LaVergne TN
LVHW022008200926
848481LV00007B/2280
* 9 7 9 8 3 4 9 5 2 1 1 0 2 *